P9-DFH-707

THE PERFECT PREDATOR

THE PERFECT PREDATOR

A SCIENTIST'S RACE TO SAVE HER HUSBAND FROM A DEADLY SUPERBUG

STEFFANIE STRATHDEE, PHD
THOMAS PATTERSON, PHD

With Teresa Barker

NEW YORK BOSTON

This book is not intended as a substitute for medical advice of physicians. The reader should regularly consult a physician in all matters relating to their health, and particularly in respect of any symptoms that may require diagnosis or medical attention.

Jacket design by Amanda Kain

Jacket image © Omikron/Getty Images

Hachette Books

Hachette Book Group
1290 Avenue of the Americas
New York, NY 10104

hachettebookgroup.com
twitter.com/hachettebooks
First Edition: February 2019

Hachette Books is a division of Hachette Book Group, Inc.

The Hachette Books name and logo are trademarks of Hachette Book Group, Inc.

The publisher is not responsible for websites (or their content) that are not owned by the publisher.

The Hachette Speakers Bureau provides a wide range of authors for speaking events. To find out more, go to www.hachettespeakersbureau.com or call (866) 376-6591.

Library of Congress Control Number: 2018960533

ISBNs: 978-0-316-41808-9 (hardcover), 978-0-316-41807-2 (ebook)

Printed in the United States of America

LSC-C

10 9 8 7 6 5 4 3 2 1

To our children,
Carly, Frances, and Cameron

CONTENTS

PART III
The Perfect Predator

PART IV
The Darwinian Dance

THE PERFECT PREDATOR

Everybody knows that pestilences have a way of recurring in the world; yet somehow we find it hard to believe in ones that crash down on our heads from a blue sky. There have been as many plagues as wars in history; yet always plagues and wars take people equally by surprise.

—*Albert Camus,* The Plague

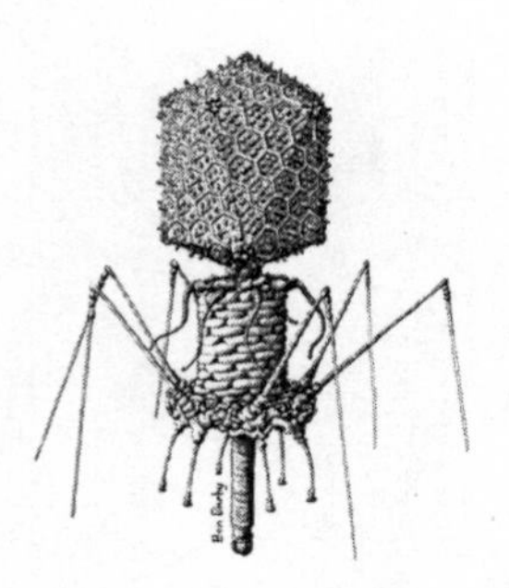

Drawing of a T4 myophage, similar to several of the bacteriophages used to treat Tom. *Drawing by Ben Darby*

PART I

A Deadly Hitchhiker

We stopped looking for monsters under our bed
when we realized that they were inside us.

—*Attributed to Charles Darwin*

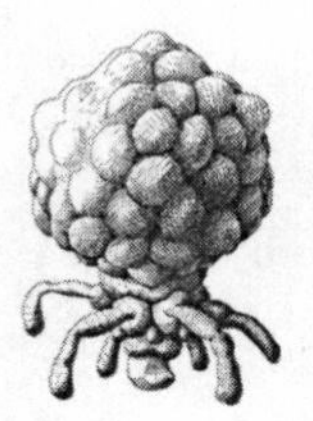

Drawing of a T7 podophage, similar to the "superkiller" bacteriophage used to treat Tom. *Drawing by Ben Darby*

BLINDSIDED

University of California–San Diego
Thornton Hospital, La Jolla
February 15, 2016

I never dreamed I'd be outwitted by a wimpy bacterium. I'd tracked a killer virus across multiple continents to wage the war against AIDS, through the trenches and at the table with policymakers at a global level. Viruses were to be feared. Bacteria? Not so much. At least not this one. I'm an infectious disease epidemiologist, director of a global health institute at a major US university, and of all people, *I* should have been able to protect my husband from a bacterium I'd last seen in my undergrad days, when we'd handled it without concern in basic lab experiments. If someone had told me that one day this microbial mutant would have us on death watch and I'd soon be injecting my husband with a legion of killer viruses to try to save him, I would have thought they'd lost their marbles. And yet, here we are.

The holidays—Thanksgiving, Christmas, New Year's, and Valentine's Day—have passed in a blur. Tom is hardly recognizable beneath the web of IVs, monitor cables, drains, tubes, and other medical paraphernalia. His once thick silver hair, which stylists swooned over, has fallen out in clumps, and the skin on his feet and hands is peeling off in layers. He has lost more than a hundred pounds from his six-foot-five-inch frame. We have not lost hope, and on this day, like every day, we are strategizing how to beat this thing. But at this moment I am doing it on my own. Tom is lapsing in and out of consciousness, an improvement over the coma, but still...

The tone of the clinical conversation among the specialists and other medical staff around Tom has changed in some subtle way. It's hard to nail down. His labs and vital signs fluctuate as they have for three months now, so it's not that. It's something between the lines, something they're *not* saying, that I'm unable to decipher. Since our lives went from bliss to hell in a handbasket, it's been all I could do to learn enough about anatomy and medicine just to keep up with their conversation. I'm a researcher, not a doctor, but even I know something about bedside manner. And theirs has shifted.

Now, the doctors and nurses speak in hushed tones and some seem afraid to look at me. In short snippets, between running exchanges with doctors and hospital staff, I turn to the internet, where I enter phrases like "alternate treatments" and "multi-drug-resistant bacteria" into PubMed, a search engine beloved by scientists. Ordinarily, my online searches are specific and hyperfocused because I usually know what I am looking for—like "prevention" and "HIV transmission" and "injection drug use." But right now, I'm not so much an epidemiologist as I am the wife of a very sick man. I'm not sure what the operative questions should be, or what a useful answer might look like. And what's freaking me out is that none of the docs treating Tom seem to know, either.

What's obvious at first glance in the scientific literature just confirms what we already know: Tom is up against, as one study says, "a difficult-to-treat pathogen whose antibiotic resistance patterns result in significant challenges for the clinician." *No shit, Sherlock.* What we've got here is one of the most lethal bacteria known to humankind, a "superbug," that has mutated to resist all existing antibiotics. Recent advances in exploratory research on how to fight this superbug have all been experimental, meaning that there was insufficient data to prove that they worked, so none were approved for general use, leaving Tom's docs at a dead end in their hunt for approved treatment options. Among novel ideas out there was an approach I vaguely remember studying briefly as an undergrad—the use of viruses that prey on bacteria—but that idea appears to be nothing more than a footnote in the margins of modern medicine.

Tom lies motionless, the steady hum and beeps of monitors the only sign of life, and I try to distract myself, emailing our graduate students about their latest papers from the corner of his room. In my busy mode, trying to keep at least a nominal tether to the real world, I dial in to a conference call to join my senior colleagues on a university retreat in San Francisco. I was supposed to be there, too. But in the months since the war against pandemics took a personal turn, everyone we know has heard what Tom and I are up against and where we're holed up. Several of my colleagues ask how Tom is doing. I give them the latest rundown before telling them that I have to ring off. We say our goodbyes, and as I get ready to hang up, the chair of the meeting, a retired surgeon and former university chancellor, asks a question quietly to my colleagues, thinking I'm no longer on the phone.

"Has anyone told Steff that her husband is going to die?"

1

A MENACING AIR

12 weeks earlier
November 23–27, 2015

It had all started out so *ordinary*. Well, ordinary for a couple of globe-trotting scientists who go *looking* for trouble in the world of infectious diseases.

Egypt hadn't seemed like a dangerous place to go when we'd started planning our dream vacation, but a month before we left, a plane was blown up near Sharm El Sheikh, Egypt's famous beach community. A few weeks later, a series of orchestrated terrorist attacks in France shook Europe to its core, with blame cast on extremists in the Middle East and Northern Africa. The tourist industry in Egypt took a nosedive. By Tom's logic, it was the perfect time to go.

Given the circumstances, I had suggested to Tom a few times that we cancel, but we'd both just launched several new research projects and were desperate for some downtime. Admittedly, our threshold for risk is a little higher than the average Joe. Our research on risk factors for HIV, sex work, and drug use routinely takes us to places where disease and street violence are everyday facts of life—and death—for many people. For thirty years before that, Tom's field work in evolutionary biology had taken him to some of the most remote places on earth, where humans weren't the favored species for survival.

As a "second time around" couple now married eleven years, our kids were grown. We were empty nesters with a passion for travel. We'd been to more than fifty countries between the two of us, frequently presenting our research at an international conference and then tacking on a few personal days for a getaway. Our adventures often included unexpected challenges. We'd fended off a rogue hippo from a dugout canoe on the Zambezi, and in India, glistening ribbons of leeches in Kerala and giant jumping spiders in Orissa. We'd narrowly missed a terrorist attack in Mumbai and a violent coup in Timbuktu, and encountered drug cartel henchmen and police on the take in our field research. We'd long since accepted that our work came with risks, and travel did, too. That was the fun part.

Tom, in particular, had a deep interest in ancient Egyptian history, art, and culture. He had wanted to do this trip so badly for so long, and we'd had to scuttle plans before. So our excitement at finally getting to go eclipsed any other considerations. But after the attacks on Sharm El Sheikh and Paris, even well-traveled friends had raised their eyebrows when we told them we were passing up their invitations of Thanksgiving dinner for the pyramids. My parents, who had flown in from Toronto to house-sit, were a tad more outspoken.

"Bad things come in threes," my mother cautioned, as she finished chopping a fennel bulb for our dinner salad, paused to play Candy Crush Saga on her iPad, then resumed chef's duty. "Hope number three doesn't happen in Egypt," she said ominously, motioning with the point of her knife to the CNN footage of the Paris attacks.

Spontaneity had always been a common denominator for us in our travels, and we liked to meet nature on its own terms. Eleven years earlier, Tom had proposed to me as we walked along the beach in Del Mar during the bioluminescent tide, when a particular species of phytoplankton glows blue-green as it's roughed up in the surf. The shimmering ripple kissed the beach and made our footprints glow, a nice romantic touch. The thing is, this kind of bioluminescence is also a marker, a warning of sorts, of the bacteria that lurks invisibly below it, causing a toxic algae bloom known as the red tide. That pretty much summed up our life view as a couple: bask in the glow, and deal with what lurks below when the time comes. Work hard, play hard.

We were married later that year by our children in an offbeat civil ceremony at a beach house in Hawaii. At the time, Tom's daughters, Carly and Frances, were twenty-one and seventeen, respectively. My son Cameron was twelve. I bought leis, grass skirts, and coconut shell bra tops for the girls. When Cameron pouted that he didn't have a wedding outfit, I bought him a set, too. Back then, he'd sulked about his dad and me divorcing, and he seemed noncommittal about my plans to marry again. Thankfully, when we overheard his side of a phone conversation with a schoolmate, he'd said, "Yeah, I am going to have two sisters, but they're cool; one even has dreadlocks!" Carly, who had yet to shear her shoulder-length dreadlocks, had become a mail-order minister through the Universal Life Church. When I teased her about it, I learned that Tom had been similarly ordained decades before in one of his repeated attempts to dodge the Vietnam War. As Carly presided over our wedding ceremony with Cameron and Frances standing solemnly on either side of us, hands clasped and grass skirts swaying, Tom and I clinked Champagne glasses. For a couple who, as individuals, had grown from humble origins and through hard times in a number of other ways, we felt we'd led a charmed life ever since.

That night, as we finished packing to leave for Egypt in the morning, I set out keys to the house and car; directions for tending the cat, the garden, the bird feeders, and the worm composter; and instructions to the remote controls for the TV. Then I did something I had never done before. At the last minute, I penned a single page of notes that began with the words "In the event of our death..." Tom rolled his eyes skyward, but he scribbled his signature on the bottom beside mine. I read it again once through and placed it neatly on the kitchen island alongside the car keys.

••••••••

Taking the advice of friends, we'd found a respected Egyptologist to be our guide, trading our improvisational habit for a more intentional learning experience. An uneventful flight left us eager to start the day's itinerary when we met Khalid the next morning. A lithe, petite man who looked to be around forty, Khalid wore khakis, a button-down plaid shirt, and worn but very well polished shoes. He approached us with an outstretched hand and a warm smile to welcome us to Cairo. Khalid had been a guide for several documentary crews and scholarly expeditions, and he would be ours for the next week through an ambitious itinerary of pyramids, temples, tombs, and other ancient sites.

Each morning Khalid picked us up and, as we drove into the desert or hiked through ruins, spoon-fed us a condensed version of thousands of years of Egyptology, archeology, and mythology. In his stories, he wove together tales of the kings and pharaohs, their tombs, and the pyramids in a lively narrative that merged bricks and mortar with ancient mysticism. So it was that each day we were immersed in a realm where the grand and ancient architecture and liturgy of death came alive for us, from mummies frozen in time to the fantastical hieroglyphics that told their tales—the graphic novels of their time.

On what was Thanksgiving Day back home, we toured several ancient ruins within a few hours from Cairo, Egypt's modern-day capital and vibrant metropolis home to more than seven million people. Virtually all of these sites were what Egyptologists call a necropolis. In other words, mass burial graves.

Inside one museum, Khalid showed us hieroglyphics painted on a sarcophagus that outlined the process of mummification. They included multiple depictions of Anubis, who had the body of a man and the head of a jackal. Anubis was the Egyptian god of the dead, responsible for presiding over embalming and mummifying bodies, ushering souls to the afterlife, and protecting their graves from robbers and demons. Also depicted were intricate, oddly shaped tools for the task of preparing bodies for the next realm. Tom shuddered at the sight of them, never one for gruesome clinical detail, even when it's five thousand years old. I found them fascinating.

"What's that one for?" I asked Khalid, pointing to a strange little hook that a thousand-year-old dentist might wield to pull an errant tooth. Khalid looked up at us with a hint of a grin.

"That one's for removing people's brains," he replied. "Through their nose. To preserve the skull. Without a preserved body, the ancient Egyptians believed there would be no home for the soul in the afterlife, and its ghost would be forced to haunt its family members."

If the ancient Egyptians seemed to have been overly preoccupied with death and funeral preparation, it was because they believed the spirit faced a perilous journey to reach the afterlife and settle in for eternity. I could appreciate their desire for a smooth journey and their abundance of caution to ensure it. Tom might scoff at my thorough notes for house sitters, but as an evolutionary biologist he had to appreciate that it's only human.

About an hour south of Cairo, we made our way to the royal desert necropolis of Dahshur and the Red Pyramid, the largest of three pyramids neighboring a military base. Because of its proximity to the base, the pyramid had been closed to tourists for many years, and even now was not always open. But we'd lucked out; it was open today and there was no one else in sight. Khalid said it was possible to crawl inside the tunnels for a close-up view. Ignoring the heat and swirling red dust, Tom and I raced to see who would be first to explore the pyramid. To reach the entrance we first had to climb several steep staircases that zigzagged up the face of the pyramid to a makeshift door, several hundred feet above the ground. Tom is nineteen years my senior, but you wouldn't know that from looking at him. Tall, with broad shoulders and a natural athleticism, he'd always been fit—a diehard surfer undaunted by the risk of "going over the falls" in "wave architecture" with dubious descriptors like "gnarly" and "walled up." From lanky youth to a sturdier middle age, Tom was always among the first to get into the water when storms or wild waves made retreat a reasonable option. He rarely considered retreat a reasonable option, period.

Using his long legs now to his advantage, Tom took the stairs two at a time and, despite having added a bit more heft to "sturdy" in recent years, was first to reach the small platform at the pyramid door. He gloated over me when I arrived a few minutes later, huffing and puffing. A lone watchman, an aging militiaman in worn army fatigues, crouched by the low entrance door, observing our arrival with cool interest. His head wrapped in a turban, with one hand he absently stroked his long white beard. In the other he clutched a battered AK-47, which lay lazily across his lap. I peered past him. Khalid hadn't exaggerated when he'd said you can crawl inside. More precisely, you *had* to crawl to get inside. I took a deep breath, lowered my head, and stepped down the first few rungs, but that was enough. I scrambled up, and Tom gave a smug smile as he started a backward crawl downward into the pyramid.

"Do not breathe the air!" the watchman called down to Tom. Local lore had it that noxious gases lingered in the chambers. Tom scoffed. Poisonous gases? It sounded like a line fed to gullible tourists.

"Famous last words," Tom hollered back. He quickly disappeared from view into the bowels of the Red Pyramid, the top of his silver-haired head no longer visible. I swallowed hard, and tried to shake a growing feeling of dread. Despite the heat, I shivered. The guard's leering presence made me uncomfortable. I turned to face the desert, scanning the dunes for Khalid. He was a black dot on the horizon. I bellowed down the chute at Tom. "Hurry up!"

Tom finally emerged, panting, covered with sweat and red dust, his face a little ashen. I handed him a bottle of water from my pack.

"Let's get out of here," I said, tugging his shirtsleeve. Another tomb awaited.

A short drive from Dahshur was Saqqara, the main necropolis for the ancient Egyptian capital of Memphis and onward for over three thousand years, well into the Roman Empire. Tom was still looking a little bushed from our last hike when we arrived at the Step Pyramid. He wandered out among the alabaster benches that lined a catacomb of tombs, closed his eyes, and took a deep breath. I could see that his brow was now coated in sweat and he was breathing more heavily, as if we were climbing a mountain.

"You okay?"

He waved off my concern.

"There's just something about this place," he murmured absently. "It's strangely familiar." We both knew he had never been to Egypt before, and Tom was skeptical about reincarnation.

"Maybe just creepy," I said.

Khalid caught up with us and resumed his informative lecture as we walked through the ruins of the necropolis; below us were the catacombs, the subterranean crypts, with mummies and the king's chambers. During the king's rule, no effort was spared to keep his larder and spirit stoked, to protect his *ka*, or his energy, his inner power. If his *ka* weakened, Khalid explained, he could be deposed by enemies. Saqqara had also been an important destination for Egyptian pilgrims who belonged to several different sects. Recent excavation had uncovered almost eight million animal mummies, including dogs, cats, baboons, falcons, and ibis.

A glimmer of interest broke the pall on Tom's face.

"Hey, Dr. Doolittle," I quipped, using one of my favorite nicknames for him. "Maybe you wrapped animal mummies in another life."

••••••••

From the time Tom was young, he'd shown a special affinity for animals. He considered this a natural inheritance from his great-grandfather, who was supposedly Cherokee. In Tom, it developed into a passionate vocation, and later a scholarly one, as he first pursued research and training as a primatologist, publishing studies on the psychology and memory of lowland gorillas. Later, he became an ornithologist, studying how the dialects of white-crowned sparrows had evolved differently from one part of San Francisco to another, publishing his dissertation paper in one of the world's top journals, *Science*. He liked to tell people that he later worked his way down the evolutionary ladder to study humans. When we went on our usual walk at the local lagoon, Tom was more likely to say hello to other people's dogs than their owners.

Eventually his career evolved to focus on long-form history—the evolutionary origins of behavior. He took the very, very long view of the natural world and how we—and all creatures great and small—evolve to adapt to a changing environment. Or perish. It's our ability to adapt in response to the stressors around us that helps us cope and survive. With that long view as his preferred one, Tom considered the world not in terms of days, months, or even years, but in millennia. He'd point to the topknot of some obscure bird species in Africa relative to a comparable one in Asia, musing aloud how they reflected "convergent evolution," the way different species develop analogous traits as they adapt to similar environments. Rather than feel sorry for himself if he got the flu or even when he was infected with a voracious parasite he picked up in the Colombian rainforest, he'd marvel at the biological brilliance that enables an organism to outwit challenges to its survival, and the adaptive value of surviving it. He was annoyingly quick to remind me of the evolutionary bright side: "What doesn't kill you, *actually does* make you stronger."

That afternoon at Saqqara, Tom was evolving right before my eyes. Within seconds, he seemed a thousand years older, his face pale and drawn. His *ka* was definitely weak. Even so, he remained intent on exploring the underworld of the ancient necropolis. Khalid and I had to tear him away.

"Just the heat," he said. By morning he seemed rested and ready to go again. We forged ahead by car, camel, and air and on foot, with Khalid in the lead, to see the temples of Ramesses II and Nefertari, and the Aswan Dam. Finally, we boarded the MS *Mayfair*, a cruise ship that would take us to Luxor for the grand finale: the Valley of the Kings. The ship accommodated 155 people and ordinarily would have been fully booked this time of year. But the recent terrorist scares and drop in tourism had left it eerily vacant. We didn't mind. We'd looked forward to some special downtime together, and an empty cruise ship certainly promised that.

As we'd learned from Khalid, Egyptian mythology held that at the end of each day, Ra, the sun god, descended into the underworld on his solar boat, where he encountered demons and gods who opposed him, such as Apep, often called Apophys, the god of chaos. Apophys would partially swallow Ra, leading to the setting of the sun each night, and would spit him back out at dawn. We'd had it easier, returning to our cozy accommodations each evening and updating our friends and family on Facebook before calling it quits.

When our ship anchored in Luxor the next day, there were several other ships but not enough docks. Our crew tied the *Mayfair* to a ship that had already moored, which in turn had tied itself to another, and so on. Hand in hand, Tom and I traversed three ships to get to shore so that Khalid could take us to visit the Luxor and Karnak temples. We returned to the ship at dusk for a romantic dinner on the top deck under the stars—a towering platter of seafood paella and a bottle of wine that I had saved for the occasion. In the days and weeks ahead, I would come to refer to our feast under the night sky as the Last Supper.

2

THE LAST SUPPER

November 28, 2015

We hadn't planned on having the ship to ourselves, nor could we have known the evening would be so spectacularly beautiful on that upper deck, under the blanket of stars and the warm, velvet breeze. But we had planned on a special dinner, and I'd packed the special bottle of Chardonnay from home and asked the ship's chef if they could chill it to go with our meal. We were celebrating the last night of this dream vacation, after all. Plus, it was the anniversary of our first date fourteen years before, and, as is my habit, I rarely left anything to chance. Organization, focus, and follow-through are the Day-Glo genes in my DNA. Tom was an amused but graceful beneficiary. If his "never retreat" ethos was a signature trait, then mine was "never give up." It had steeled me through a childhood as a brainy, bullied kid, and earned me the nickname Pit Bull in my early academic career.

Tom sometimes chafed at my dogged persistence and attention to detail over things that he might have preferred to put off for another day. But he also mused, affectionately at times like this, how the forces of nature and nurture that were so different for each of us could have forged the likes of us as a couple. Tom had grown up something of an Oliver Twist in 1950s Southern California. I'd grown up in the seventies white-bread middle-class Toronto suburb of Scarborough, with two sisters and a mother and father who provided for us in all the ordinary ways a kid can take for granted. Our job was to study hard and get good grades.

My dad was a high school science teacher who later took on mentoring gifted kids, and I guess he'd practiced on me from the start. I was the nerdy girl who took even that a step beyond the norm. Today I'd be described as a little "on the spectrum"—a euphemism for Asperger's syndrome, though I've never had a formal diagnosis. With a high-octane brain for academics but low-res skills for connecting with kids my own age, I was forever on the fringes socially, but had a couple of girlfriends who shared similar interests. Meanwhile, my dad tutored me painstakingly in physics and math, showing me techniques for testing a hypothesis to determine the "knowns" from the "unknowns"—the analytical heart of scientific enquiry. I was required to study two hours after dinner, and my sisters and I were only allowed one hour of TV per week. When I told my dad that I'd decided I wanted to be a scientist, he thought it was because I was trying to follow in his footsteps, but the truth was I wanted to prove to myself that I could do something really challenging. Once I discovered a passion for problem-solving science puzzles, I'd found my path.

Tom had the same drive to tackle knotty research problems, although we'd each traveled markedly different paths in our pursuit. My trajectory was very linear. I'd gone right into college and then grad school, became a postdoc, and then faculty. Tom's path was a little more circuitous in that he'd embarked on his postdoc work and then worked as a high school teacher for three years before returning to academia full time after he obtained research grants. Our paths would eventually intersect, but even before then our research on the frontier of the HIV/AIDS epidemic slowly brought us into the same orbit of academic publishing and international conferences.

By then, Tom had already turned his attention to humans rather than animals and fish; his focus was on stress research. He'd started off studying elderly people and what made them vulnerable to illness, which led him to the study of psychoneuroimmunology—the effects of the mind and brain on health and the immune system. When the HIV epidemic hit, he was riveted by the potential to study the far-reaching implications of stress on the immune system, and a person's vulnerability and resilience in the context of this disease. He became fascinated that some people progressed to AIDS and others didn't, and he wondered why—this was one research question we had both studied. He was awarded a grant to study the impact of stress on HIV disease progression, which led him to study risk behaviors that made people vulnerable to infection. He was one of the first people in the field to develop a risk-reduction program for people who were already HIV-positive—recognizing that people needed skills to help prevent them from passing the infection on to others.

Each on our own path for the previous twenty years, we had been following the same north star in our research, mapping the constellation of behavioral risk factors driving HIV infection. When we finally met face-to-face in 2001, it was at a boring grant review panel. Tom was developing risk-reduction programs for methamphetamine users and had just started working in Mexico with sex workers. I had just returned from Pakistan, where I was studying the impact of the post-9/11 war in Afghanistan on the risk of acquiring HIV among people who inject drugs. We often joked that of all the most dry, unromantic, and unlikely places to discover the love of your life—personal chemistry ablaze—this scientific committee meeting would have topped the list.

It would be years later, long after we were married, that we realized we'd actually first "met"—eyes only—across a crowded room at a conference in 1997 in Flagstaff, Arizona, where I'd presented my most recent research findings. It was a riot because, chatting one day about a new study that had just come out, I referred to something about that conference talk, and Tom said, "Wait I minute, I was there, too! Were you wearing a yellow suit?!"

I had to stop and think. I'd worn a crisp, tailored yellow suit and stiletto heels, remarkably more attentive to looking polished and "grown up" at that young age than I would become in later years. And now I remembered the lumberjack in the back row.

"Yes! Were you in the back row wearing a red plaid shirt?"

"Yup, but you walked right past me," Tom said, feigning insult.

"Yeah, I wasn't ready for you back then," I said, laughing.

"Neither was I," he quipped. And we were right.

A few years later when we were dating, we were visiting Chicago, having dinner in the Signature Room on the ninety-fifth floor of the John Hancock Center. We had just ordered and were looking out the window at the view, when *BAM!* Out of nowhere, a peregrine falcon plunged down onto a pigeon, and feathers went *flying*. We both blurted out, "Did you see that?!" And then looked around. No one else had. Just us. We proceeded to compare birding notes. Tom, who had spent years crawling around Twin Peaks in San Francisco looking for the nests of the white-crowned sparrow for his dissertation, immediately speculated that the peregrine had a nest around there somewhere. I could tell he was itching to peer out the glass to see if he could find it. That's when I knew I was in love. Now, at home, a pair of peregrines lived nearby and regularly rocketed through the canyon behind our backyard.

Toasting the luck of cosmic timing and life's mysterious ways, we turned to the chef's stunning seafood feast, the wine, the Champagne, and the silken flan. Even I had to admit that it couldn't have been a more perfect anniversary dinner if I'd planned it all myself. By the time we trundled back to our cabin, we were relaxed, romantic, and looking forward to the next day's grand finale, the tour of the Valley of the Kings, and our flight home.

••••••••

I felt the bed shake and looked up to see Tom throw the covers back and dash for the bathroom. I squinted at my phone to see the time: midnight. Tom barely made it to the tiny toilet in our cabin, where he vomited every last morsel. Or so it seemed, but his misery continued. He had terrible stomach pains and made countless repeat trips to the loo that night. Neither of us slept. When Khalid appeared in the morning at our door, Tom was doubled over in the bathroom praying to the porcelain god. Khalid urged us to call a doctor, but Tom was dead set against it, and I didn't push. Maybe he'd simply gotten a bad mussel and needed to ride out a bout of food poisoning. We were old hands at that.

Tom made feeble attempts at humor. "Did I inhale too much of the poisonous gases in the Red Pyramid the other day?" If he could laugh, I took that as a good sign.

"I just told everyone on Facebook that at this rate, you're going to see the Valley of the Kings feet first in your own tomb," I joked. Tom didn't laugh this time. He moaned, through clenched teeth.

"Just get us home *now*."

Home? That seemed a bit extreme. We'd run into things like this before in our travels. Delhi belly, we called it—our name for traveler's diarrhea or stomach upsets from unfamiliar microbes in local food and water. We routinely packed Cipro, an antibiotic, like some people pack toothpaste. And where was my guy who typically pooh-poohed physical discomfort, braved the worst waves, and didn't consider retreat a reasonable option? I was less sympathetic than I might have been. I contemplated calling a doctor or finding an emergency clinic in town where they'd check him out, give him a prescription, and in twenty-four hours he'd be good.

"Don't call a doctor," Tom snapped, reading my mind. "And whatever you do, don't take me to a hospital. I won't go."

Tom had a history of avoiding doctors. His hardscrabble childhood had trained him for stoicism. His dad, a decorated World War II Navy veteran turned motorcycle cop, kept the life lessons simple: whatever happens, you tough it out. Once, when Tom was a teenager, his surfboard boomeranged back and hit him square in the face, breaking teeth and bone. When he returned home and showed his dad the blood gushing from his mouth, his dad shrugged, pried the gash open, and poured in mercurochrome, that god-awful, glow-in-the-dark concoction that parents favored as a cure-all antiseptic back then. When Tom went to the dentist months later, the dentist rocked back on his heels, gave an amazed whistle, and asked Tom how long ago he had broken his jaw and why he didn't have it properly set. I'd been raised to take care of myself, too, but with parents who had a more conventional relationship with modern medicine.

Tom and I agreed on almost everything, but when we didn't agree, we locked horns. Neither of us knew how to give in. I was determined that now would not be one of those times. But Tom was sick, and I was not. He wasn't thinking clearly, so I had to. He tended to trust me on the medical side of things, since my degree in epidemiology was a branch of public health. But having a PhD without an MD meant that sometimes I knew just enough about medicine to get me into trouble.

3

DISEASE DETECTIVES

Epidemiologists study the big picture: patterns of disease, how they spread, and how to stop them, whether through biomedical means or changing people's risky behaviors. My mother often described what I do for a living by saying, "You know when they say that some disease is spreading or an epidemic is on the way? Well, my daughter is one of the 'theys.'"

Not that I'd started out to be one of them. I'd been an undergraduate major in microbiology at the University of Toronto when Rock Hudson died in 1985 of a mysterious illness that was first referred to as "gay-related immunodeficiency"—or GRID. My professor, Stan Read, was an early leader in the field and I learned the hallmark signs of the infection—night sweats, swollen glands, and flu-like symptoms, sometimes accompanied by other infections typically seen in people whose immune systems were suppressed. The first cluster of cases of what is now called acquired immunodeficiency syndrome (AIDS) had emerged in 1981 in gay men, capturing the attention of epidemiologists at the US Centers for Disease Control (CDC), who conducted a series of outbreak investigations. As further research in North America and Europe found that hemophiliacs and injection drug users were also succumbing, it was an inspiring time to become a disease detective. Within a few years, a diagnostic test was developed based on antibodies directed against the causative virus, human immunodeficiency virus (HIV), and I'd found my calling, or so I thought.

But a few summer internships fumbling under the flow hood had convinced me that day-to-day lab work was not my forte. Then a graduate teaching assistant suggested that because of my preference for working with people rather than Petri plates, epidemiology might be a perfect fit: all the intrinsic beauty of biology, but instead of being about populations of cells, it was about populations of people. Epidemiology was the big-picture, bird's-eye lens needed to develop strategic action to help people. When the first HIV cases appeared in my hometown of Toronto, I was moved by the drama unfolding before my eyes. I obtained my master's degree and started my PhD. I was to become one of the few HIV epidemiologists in Canada at the time. Eager to help while still a grad student, I was one of the first group of volunteers at Casey House, which started out as one of the earliest dedicated AIDS hospices in North America. There I watched a growing number of people who had become my friends die painful, lonely deaths. In the early nineties, my doctoral supervisor, Dr. Randy Coates, and then my best friend, Michael, both of whom were stricken with HIV, died within a year of each other from opportunistic infections that they would have been able to beat if their immune systems hadn't been annihilated. I was devastated. And that was it: I became committed to studying this pandemic and finding ways to stop it. That was the first time my personal life and my professional life had collided, but it wouldn't be the last.

During that same period, I learned how, even as scientists, our intuitions can help us. When I was working on my master's degree, I'd worked part-time as a research interviewer for an HIV study on people who inject drugs and sell sex. The participants with the highest-risk behaviors often confided that they had been victims of child sexual abuse, even though that wasn't an interview question. I told the principal investigator about it, but at that point she couldn't change the study questionnaire, which meant that we couldn't study it. But I never forgot about those stories. Years later, when I was directing studies of people who inject drugs and young gay men in Vancouver, I included questions on childhood sexual abuse and discovered that it was strongly predictive of HIV risk behaviors in both populations. My finding became part of a budding body of research that showed that social determinants—including the driving forces in our economics, politics, and laws—helped explain the marginalization and stigma in these key populations, increasing their vulnerability to HIV. I received a Young Investigators Award at the 1996 International AIDS Conference for that research, and that early insight inspired my career path to shift away from a focus on individual behavior change to change the conditions that put people at risk in the first place. Beyond that recognition was the power of the lesson I'd learned as a young scientist and practiced ever since: listen to your intuitive hunches, don't let the unknowns limit your search, and approach all possibilities with an open mind and rigorous research.

At the moment, my intuitive antenna wasn't picking up much from the bathroom, where Tom hunched in pain. Then the Red Pyramid guard's ominous warning about the "poisonous gases" in the tomb echoed in my mind. Until the end of the nineteenth century, most people, including scientists and physicians, embraced miasma theory, the belief that "bad air" was the cause of diseases we now know to be caused by bacteria, like bubonic plague, cholera, and syphilis. Their opponents who supported germ theory were outnumbered and sometimes even outcast. That is, until John Snow, a medical doctor who would later be known as the father of epidemiology, conducted his famous investigation into the London cholera epidemic. That story had drawn me to the field of epidemiology and was later the subject of one of my favorite books, *The Ghost Map*, which laid out his meticulous detective work.

After studying the first cholera pandemics that occurred earlier in the nineteenth century, Snow observed that they tended to follow trade and military routes. He hypothesized that those afflicted might have been exposed to water that was contaminated with some sort of microorganism. But his 1849 article on the subject was criticized because he had no proof. With the London outbreak, he got his chance. London had burgeoned into a megacity without a sewer system for handling the tons of excrement generated from two million people and their animals. The city's poop was collected by so-called night soil men, who carted it off to cesspools in the boonies that stank to high heaven. The public health commissioner, a fervent believer in the miasma theory, was intent on creating a sanitation system to rid the city of its smelly cesspools, which now numbered about three hundred thousand. There was one problem. He was focused on cleaning the air, not the water. His solution was to eliminate the cesspools and dump the masses of poop into the Thames River, which turned it into a polluted mess that inadvertently contributed to cholera outbreaks that would kill thousands.

The 1854 outbreak began quietly but ferociously in London's Soho district, which was one of the city's poorest. Within days, hundreds then thousands of people were overcome with terrible stomach pains, explosive greenish diarrhea, and extreme dehydration that more often than not quickly led to their deaths. John Snow, who lived in a neighboring district, took it upon himself to investigate, putting himself at considerable risk due to the neighborhood's high crime rate, in addition to the risk of acquiring the infection himself.

Suspecting that a water source was the culprit, he sampled water from the closest pump on Broad Street to inspect under a light microscope. But microscopy at that time fell short of what Snow needed to find conclusive evidence. The offending bacterium, *Vibrio cholerae*, was actually isolated that same year by Italian scientist Filippo Pacini, but it would take decades before its significance was recognized. Meanwhile, the cause of the Soho cholera outbreak remained elusive until Snow literally took a broader view. He knocked on doors to survey people about where they drew their water, tracked which pumps were linked to the city's sewer system, and then which of those drew water from the Thames. Comparing his data and death statistics from adjacent neighborhoods, he showed that cholera death rates differed dramatically based on their water source. With that evidence, he convinced the Board of Health that the Broad Street pump was the point source of the outbreak. In a symbolic gesture that would become a critical moment in medical history, the Broad Street pump handle was removed, ushering in a new era in which the miasma theory was eventually abandoned in favor of germ theory. The field of epidemiology was born.

••••••••

Whether you're tracking the cause of cholera, HIV, or some other outbreak of disease, the big breakthroughs often come from simple, dogged detective work—what a professor friend once called "gumshoe epidemiology." In the middle of the night, in bed on a cruise ship overlooking the Nile, simple was the only kind of microbial detective work possible. My undergrad years in microbiology gave me a place to start.

First, I considered what we had eaten: fish, clams, prawns, and mussels. Shellfish are often contaminated with fecal coliforms, which are bacteria found in human and animal poop. So we could be looking at a virulent strain of *Escherichia coli* or any number of *Shigella*, *Salmonella*, or *Vibrio* species. On my smartphone, I googled a few of the more exotic bugs and added *Listeria* and *Clostridium*.

Next, I did a simple calculation to take into account that bacteria have different incubation periods depending on the time of exposure. When had we eaten relative to Tom's first onset of symptoms? It was past midnight now, which meant about five hours since dinner. But the culprit in food poisoning may be something you ate anywhere from an hour to two weeks or more before, depending on the organism. I googled the CDC website that tabulated incubation periods for foodborne pathogens. Scanning the list, I excluded some bugs with longer periods, like *Campylobacter jejuni* and *Vibrio cholerae*, but that still left a slew of nasties.

Tom crawled out of the bathroom, but before he could even get to the bed to rest, he turned in his tracks and stumbled back, retching.

Although I couldn't diagnose what he had, I jumped out of bed and rummaged through our suitcase for that don't-leave-home-without-it Cipro. It was usually a universal lifesaver, clearing up most cases of vomiting or the trots within hours. The problem was, Tom couldn't keep anything down. Not dinner. Not the mild broth that the ship's chef had sent, not hot tea, not even a sip of water. I watched as our only dose of Cipro was flushed down the toilet. An assortment of small dishes and cups, their contents untouched, sat like sacred offerings on the bedside table.

Could Tom have acquired a viral or parasitic infection from what we ate? We had both been vaccinated against hepatitis A, so I crossed that one off the list. Most parasitic infections would have taken longer to make him sick, so I discounted them too, at least for now. Of the possible viruses, norovirus came to mind, since it causes stomach pain and vomiting, and was the cause of most outbreaks of foodborne illnesses on cruise ships. But we were the only passengers on the ship, and neither Khalid, myself, nor any of the crew were sick.

Tom had such a cavalier attitude about food and drink on our travels that he might have picked something up anywhere in the past few days. He was no stranger to dysentery, having pushed through some pretty impressive bouts of it, so I confess I'd become a bit jaded to even that prospect. He liked to boast of the time in his younger adventuring days when he and a friend had gone off-road camping the length of Baja California, a thousand miles with no paved roads. After an initial bout of Montezuma's revenge, he figured he'd bested any bugs in the water and it wouldn't hurt to drink the water anywhere they found it. His dysentery landed him in the hospital, where the doctors identified four different pathogens. That was the bad news, but the good news was that he'd lived to tell the tale, as always. The chorus in my head was always: *By this time tomorrow, he'll be laughing at me for being a worrywart.*

Dawn came, and with it my hope that this would resolve itself. But things didn't get better. Tom still couldn't keep anything down and was growing weaker and more dehydrated by the hour.

4

FIRST RESPONDERS

I didn't want to flat-out overrule Tom on calling a doctor—even in this state he would be furious. So I did the next best thing. I texted our friend and colleague, Dr. Robert "Chip" Schooley, chief of infectious diseases back home at our institution, UC San Diego. Chip and I were both division chiefs and saw each other frequently for work purposes—sometimes socially, too, so he and Tom were well acquainted. They shared a mutually wry sense of humor, and enjoyed swapping stories from their travels. As a physician and a scientist, Chip was laser sharp and serious—qualities that distinguished him as an international leader in the field of infectious disease. At the same time, he possessed a natural warmth and wit.

Chip had come to our rescue once before, after Tom and I contracted a strange skin infection on another Thanksgiving trip we took in 2008 to Goa, in western India. By the time we arrived home, both of us had several boils on our arms and legs that started out looking like giant zits before turning into big balls of pus, throbbing and pulsing with a life of their own. None of them responded to antibiotic cream, and I immediately became alarmed.

"I bet this is MRSA," I'd told Tom. "If it is, we need to see a doctor ASAP." Tom scoffed at me.

"Isn't that just a type of *Staph*?" he'd replied as he peered at the greenish boil on his forearm. "Can't you get a stronger ointment?"

MRSA is a form of *Staph*, all right, but it's resistant to the antibiotic that used to be used to treat it: methicillin. Which is why it's called MRSA—methicillin resistant *Staphylococcus aureus*. It was the first of the antibiotic-resistant "superbugs" identified in the UK in the early 1960s, and since then it had quickly spread worldwide. It was a good bet that we'd gotten it when we took that swim in the Arabian Sea. In retrospect, we'd noted, no one else had been swimming, which should have been a tip-off.

The hotel in Goa was adjacent to a large slum, and in the distance, we could see several large oil tankers that had probably been emptying their bilges into the ocean. We hadn't given it much thought at the time. It's the ocean, right? A habitat for thriving thousands of species of sea creatures, flora and fauna. Who knew that there are 100 million times as many bacteria in the oceans as there are stars in the universe? I should have known that sea water is often contaminated with some of the gnarliest bacteria around.

MRSA had been all over the news at the time, due to a few high-profile cases that implicated this and other antibiotic-resistant bacteria in flesh-eating disease; some people died or had limbs amputated. Tom thought I was being histrionic, but with the Thanksgiving holiday around the corner, I convinced him that I should contact Chip for advice, and he had paid us a house call that afternoon. Chip looked a bit like an aging Alfred E. Neuman, the cover boy of *Mad* magazine, sans the protruding ears but with a generally jovial expression that put you at ease. I hadn't often seen him without his lab coat. Chip had just returned from a trip to Mozambique, where he had established a new medical education program, and that day he was in shorts and a T-shirt, and his freckled face took on the sheen of burnished copper.

He reached into his pocket to pull on some gloves while I pulled up the hem of my dress to show him the boil on my leg.

"Whoa, that's MRSA all right," Chip chortled. "I don't need to come any closer. Of course, to properly diagnose it, we would need to culture it. But that would take a few days, and you need some antibiotics right away. I'm betting this dip you took in the Arabian Sea gave you a superbug souvenir. The good news is that MRSA from natural environments like this is usually still sensitive to oral antibiotics. The trickier ones to treat are strains of MRSA and other multi-drug-resistant bacteria that are acquired in hospitals, because they've grown in what amounts to a hothouse for developing antibiotic resistance."

We were lucky. After two weeks of oral antibiotics, our boils dried up, although we missed Thanksgiving dinner that year because we didn't want to pass on our infection to anyone else. Within six months, our immune systems cleared the bug and we had forgotten all about it. Except when we went to the dentist, where I routinely reminded the hygienist that we had been MRSA carriers; it was a sure-fire way to avoid a deep clean.

••••••••

Fast-forward seven years, and Chip was playing first responder once again. He was en route to his project in Mozambique again and the cell phone reception wasn't great, but I got the gist from our brief conversation: call a doctor. He thought my gumshoe detective work that pointed at food poisoning sounded reasonable, but at the very least, Tom would need IV fluids. Against Tom's continued protests, I happily blamed Chip and asked Khalid to summon a doctor to the ship.

Dr. Busiri arrived within an hour in the early afternoon carrying a classic black leather medical bag and wearing a pristine white doctor's coat over a dark green button-down shirt tucked into pressed dress pants. He was tall, thin, and clean-shaven, with an air of efficiency. He'd had to make his way across the three docked ships and up a narrow three-level winding staircase to reach our room, but he didn't complain. He shook my hand and smiled warmly, then turned to Tom, who was pale and sweating in the bed. He took Tom's vitals calmly and quickly—temperature, heart rate, blood pressure—then pulled out a notepad and began to ask questions about Tom's medical history.

History of heart problems? Diabetes? No heart issues, and although

Tom had borderline diabetes, nothing would suggest a tie-in with this gut attack.

"He needs IV fluid," Dr. Busiri concluded, removing the stethoscope from his ears, and reaching for a collapsible IV pole he had brought along. "I suspect food poisoning, so I will prescribe some gentamycin. That always works." He tied a rubber tourniquet around Tom's arm, flicked a syringe to expel any air bubbles, and swabbed the vein inside the crook of Tom's elbow. Tom winced as Dr. Busiri inserted the needle. "The fluid will make him feel much better," Dr. Busiri said as he attached the IV bag to the pole. "I expect him to be up and around before dinnertime."

I sighed with relief. Dr. Busiri stayed nearly an hour, until the last drip of saline and antibiotics had drained into Tom's arm. "If he is not better, call me," he said, as I walked him to the door, paid him, and thanked him again.

Tom slept a few hours, and at dinnertime I tried to rouse him. He groaned and waved me off, wanting to sleep.

"Aren't you feeling any better at all?" I asked him. He shook his head no. I tried to coax him to drink some soup, but he wouldn't eat anything.

His stomach was a distended balloon. Over the next few hours, he continued to vomit. He hadn't eaten in nearly twenty-four hours, and this nonstop barfing seemed humanly impossible.

"My back is killing me," he whispered.

"Your back?" That was surprising—definitely not a symptom related to food poisoning. Was the bed bothering him from lying in it so long, or was this something else? Tom grimaced.

"Yeah, the pain, it's—it's like a band radiating from my stomach around to my back."

That rang a bell but I couldn't remember where I'd heard it. Later, when I closed my eyes and tried to doze, it came to me. A few months earlier, one of our cats, Madame Curie, went off her food and started vomiting. After a few days, when I picked her up to put her in her cat kennel to take her to the vet, she wailed when I touched her stomach and her back. The vet's initial diagnosis was pancreatitis, inflammation of the pancreas, although she could not pinpoint the cause. Poor little Curie died a few days later. Granted, a cat's anatomy was a far cry from a human's, but maybe Tom didn't have food poisoning. Could he have pancreatitis? I did what anyone else would do in my shoes. I googled it. Within seconds, my cell phone lit up in the dark and listed the symptoms: vomiting, stomach pain, back pain. *Shit.*

I texted Chip again.

Got a min? Tom is worse & now has back pain. Could this be pancreatitis?

Chip texted me back within five minutes and asked me to call him. After I brought him up to speed, his tone was no longer jovial.

"It could indeed be pancreatitis, or it could be a twisted bowel—the other possibilities are worse. Get him to a hospital. Make sure they take you to wherever the expats and tourists go. And Steffanie—call now. *Stat.*"

Shaken, I hung up to tell Tom he'd been overruled. He needed to get to a hospital, and there was no way I'd be able to get him down that spiral staircase, across those docked ships, and up those sloping steps in the dark to the landing and into a cab for a ride to the local hospital by myself. We needed an ambulance. But there was no calling 911 in Luxor. In fact, the town didn't even have a hospital. I rang Khalid, who in turn called Dr. Busiri. He insisted on examining Tom again before we did anything further. By the time he arrived, it was eleven thirty p.m. Dr. Busiri took Tom's vitals and frowned.

"His heart rate remains elevated and his blood pressure is dropping. We need to get him to the clinic immediately," he said to me, calmly and quietly. "Your husband is going into shock."

So was I.

5

LOST IN TRANSLATION

Luxor, Egypt
November 29–December 3, 2015

Like pallbearers on a frantic midnight mission, the eight men gripped the gurney, Tom strapped to it, as they scrambled across the three ships moored between ours and the dock, then up the ancient stone steps to the loading dock. At the top, the ambulance waited in a glow of yellow flashing lights. A clutch of onlookers stood watching the scene before them, and some peered in after the men hoisted Tom and his stretcher into the back. Khalid had helped carry Tom up, but now darted to his tour company car to meet us at the clinic. I clambered into the ambulance next to Tom and Dr. Busiri.

The roads across town to the clinic were ragged with potholes, and with each jolt, Tom howled in agony.

"Can't you give him something for the pain?" I asked Dr. Busiri.

"Not until we know what the problem is," he replied.

Just the assumption that we'd soon have a diagnosis was reassuring in and of itself. By my thinking, once you know what a problem is, you can figure out how to solve it, right?

It was one a.m. when the ambulance pulled up at the clinic. The street was mostly deserted, but it was clearly a major corridor for traffic during the day. Set in a small storefront, the reception area opened up to a series of smaller rooms and an operating theater that was accessed through a swinging set of double doors. I looked around the spartan surroundings with a sense of foreboding. Tom was wheeled into the operating theater on the stretcher, with me and Dr. Busiri close behind. The room was full of empty bed frames, a few of which were equipped with thin mattresses.

Waiting to greet us were the clinic's gastroenterologist, a radiologist, and a cardiologist, all of whom were bleary eyed, having been called to the clinic for this emergency case involving a tourist. They were all deferential to Dr. Busiri, the medical director of the clinic, which was only one month old. This makeshift space was a temporary treatment center until the full remodeling on the larger facility nearby was complete. Whether the stripped-down look of the place reflected limited resources or its temporary status wasn't clear.

The staff instantly went to work, taking Tom's vitals, hooking him up to a cardiac monitor, and drawing blood.

"Why are we in an operating room?" I asked, all the more unnerved by the sound of the fear in my own voice.

"Because this is where our cardiac monitor is," Dr. Busiri replied sheepishly. It was the only one in the clinic. "But here, he can also be carefully monitored since he is in front of the nurses' station." That was more reassuring.

"What tests are you running?"

"We are drawing blood to test for cardiac enzymes," Dr. Busiri replied. I drew a blank and it must have registered on my face. "To rule out a heart attack," he explained. "Judging by his size, he is at considerable risk." He gestured to Tom's stomach, which was even more distended than it had been just a few hours earlier. The possibility of a heart attack had not even occurred to me; inwardly I started to panic.

"But we will also test for lipase, a pancreatic enzyme that can be diagnostic of pancreatitis. We will have these tests back within the hour."

Before leaving us, he motioned for the gastroenterologist, who inserted a short, flexible length of clear tubing into Tom's right nostril.

"What's that for?" I asked.

"It's a nasogastric tube. It will prevent him from vomiting further,"

Dr. Busiri replied, as his colleague finished the brief procedure, snaking the tube down through Tom's esophagus to his stomach. Instantly, the tube began to siphon off a dark greenish brown fluid into a clear collection bag that hung by his side.

"Bile," Dr. Busiri said.

I shuddered at the sight and then saw Tom watching me. He was always the first one to make a joke in the worst of circumstances, and feeling helpless to do anything useful at all, I tried to lighten the moment.

"Hey, honey, just think of the god Anubis trying to suck out your brains through this tube." Tom's eyes widened in fear, as if he thought I was serious. With a rising sense of alarm, I realized that he *did*. He was growing delirious.

Dr. Busiri looked incredulous at my poor attempt at humor and changed the subject.

"You have health insurance, yes?"

I nodded.

"It would be good to call them."

We had each spent $36 on the university's travel insurance for this trip. I could only hope it would be enough to get us through whatever lay ahead; we had never needed to use it before. I plugged my cell phone into the wall of the operating room to charge it before making the call, but the outlet didn't work. I tried another outlet; no juice. I finally found one at the far end of the room, where I needed to prop the phone up so it would stay plugged in.

At UC San Diego, I was a global health expert—a "muckety-muck" as Chip referred to me—but the professional was about to become the personal in this resource-stretched clinic. It was easy to take so much for granted in our medical system and institutions. Now we were in a country where many of the essential resources for a medical setting—resources like reliable access to electricity or some routine meds—were not a given.

An hour passed. Tom drifted in and out of sleep, the phone and internet reception flickered in and out, and my confidence began to waver, too, as I began to grasp the seriousness of the situation and my own limitations of knowledge and experience. Finally, Dr. Busiri burst through the double doors with a triumphant exclamation. "We have the diagnosis!"

Tom's lipase levels were three times normal, which confirmed acute pancreatitis. He would still need a CT scan to rule out a bowel obstruction, Dr. Busiri explained, but that would have to wait until morning. The only CT clinic was about ten minutes away, and we'd need to wait for a cancellation to fit Tom in.

"In the meantime," Dr. Busiri said, "get some sleep."

As he stepped out, a female nurse dressed head-to-toe in traditional garb stepped in and greeted me, in Arabic. The hijab covered her head, framing her face, which heightened the contrast with her dark, charcoaled eyes. With her help, I was able to find a sheet and a pillow for Tom. I rolled up my sweatshirt in a makeshift pillow for myself. I'd be sleeping on a gurney next to Tom's. His and Hers.

Once Tom was given morphine for his pain, he fell into a deep sleep. I watched him breathe and listened to the steady beeps of the monitors, the squeak and slam of the double doors as the nurses tended to other patients, and the swish of a mop from a young man who kept the clinic looking spotless. The sounds of suffering needed no translation. It would be a long night for us all.

Periodically the nurses stepped in, took Tom's vitals, and spoke among themselves outside the room. They spoke Arabic only, and I didn't, so there was nothing I could glean from them about Tom's condition. The doctors were fluent in English, having mostly been trained in Cairo, but even so, I was discovering that the technical medical terms for what was unfolding were another foreign language. Each round of tests brought a new glossary of terms. Blood tests were done to measure biomarkers, naturally occurring chemicals in the blood, and terms like *bilirubin*, and *LFTs*—liver function tests—were a continuing thread in the conversation. *Troponin*, a cardiac enzyme, was monitored regularly. The biomarker CRP—C-reactive protein—which is a sign of inflammation, was much higher than normal. In this crash course in medicalese, the vocabulary was spelling trouble.

All of the nurses were female and wore the traditional hijab, with several of the youngest wearing shrunken sweaters on top of their flowing robes, a stylish Western accent to their wardrobe, along with high-heeled shoes. Although only their faces showed, their eyes were lined heavily with black kohl, mascara, and eyeshadow, and their lips glistened with bold red lipstick and gloss. They giggled coquettishly whenever the doctors were around and paid us little attention. Their appearance was a stark contrast to me with my bare face and dirty ponytail, wrinkled T-shirt, casual cotton skirt, and flip-flops. The *Vogue* magazine buried in my duffel was as close as I got to fashion these days.

Retreating to the corner of the operating room to avoid waking Tom, I called the travel insurance company and explained our plight. I connected them with Dr. Busiri by email so that they could negotiate payment. Next, I texted Carly and Frances back home to tell them their pops was sick but that so far, everything was under control. I hoped I was right. Carly was in her early thirties and Fran in her late twenties, so both girls had plenty of prior experience with their intrepid dad's run-ins. With any luck, Tom would be treated and we could fly home within the next day or two. His track record for luck made that a reasonable expectation, despite all appearances to the contrary. He'd seen worse and lived to tell the tales.

I closed my eyes and said a little prayer. *Please, God, please let Tom be okay. And please get us home safe and sound.* I could hear a few other patients in the clinic down the hall, groaning and retching. It seemed like everyone was calling out in pain, Tom included. He woke every few hours and gripped his belly. If I didn't request more pain meds, he wasn't offered any. In the States, plentiful supply and routine overprescribing had helped create the opioid crisis. Here, they had to be frugal to a fault. I tried to limit my requests, but Tom's pain was worsening.

In the morning, Dr. Busiri paid us a visit. He looked like he hadn't slept, either. He had managed to work out the payment with the travel insurance, which was a relief. He examined Tom and started him on a new round of antibiotics, which he described as a third-generation cephalosporin. I knew this drug was a riff on penicillin, the first antibiotic, discovered in 1928 by Scottish scientist Alexander Fleming. It happened to be one of my favorite stories in the history of science: how Fleming made the landmark discovery that would transform how bacterial infections would be treated in the twentieth century and forever, the world over.

•••••••••

Fleming was already a respected scientist in 1928, but cleanliness was not his strong suit. I could relate. I had abandoned my early dream of being a microbiologist because, as a summer student, I kept contaminating my lab samples. In his lab, Fleming kept some Petri dishes filled with agar, a bench scientist's version of Jell-O, and an ideal medium for growing bacterial cultures. According to Fleming's account, he left the lids off when he went away on vacation, and when he returned a few weeks later, one of the Petri dishes was pocked with a hairy greenish fuzz. Mold. Most scientists would have just thrown the plates out, reminding themselves to be a little tidier but missing their significance. But Fleming was a keen observer and noticed something unusual. On the agar around the green mold was a clear zone where no bacteria were growing; he remarked later that it looked like the bacterial colonies nearby were dissolving. This discovery formed the basis for his paper on the inhibitory effects of *Penicillium notatum* mold on so-called Gram-positive bacteria, published in 1929, naming the active agent *penicillin*. Gram-positive bacteria, named for the scientist who created a test to classify them, include the *staphylococci* (staph), *streptococci* (strep), *Bacillus anthracis* (anthrax), and the bacteria responsible for diphtheria (*Cornynebacterium*)—some of the deadliest bacteria of the time. (Gram-negative bacteria include *E. coli*, *Salmonella*, *Shigella*, and *Legionella*.)

You'd think a major discovery like Fleming's would have had the drug developers tripping over themselves to scale up penicillin for the masses. But he had a hard time getting chemists interested in working on his strange "mold juice," which was proving difficult to produce as a stable, consistent product that could be manufactured for wide-scale use—and profit. It was a classic case of the difficulty in taking novel scientific findings from bench to bedside, and a few years later, he gave up trying. Not having been the kind of guy to toot his own horn, his paper was scarcely noticed for nearly a decade. Meanwhile, millions of people were dying from what were later considered to be run-of-the-mill bacterial infections. Among them was my great-grandmother, who died of appendicitis around 1930, when my grandma was still a schoolgirl.

By the late 1930s, efforts to isolate and purify penicillin were finally taken up by several other scientists, including Drs. Howard Florey, Ernst Chain, and Norman Heatley at Oxford University in Cambridge. They worked under austere conditions during the heart of World War II, growing penicillin in the likes of metal tins, tubs, and bedpans under the constant threat of Nazi bombing raids. Once it was realized that penicillin could potentially cure abscesses, gas gangrene, tetanus, and diphtheria, the mold was treated as a war secret. So dire was the threat of losing their potential miracle drug to the enemy or having the lab blown up in a *blitzkrieg* that Heatley thought of an ingenious way to save the culture: they rubbed the mold on their lab coats so that its spores could be recovered and recultured if necessary.

After Florey and Heatley absconded to the US to seek help manufacturing penicillin, it was used to treat Anne Miller, a thirty-three-year-old woman in Boston dying of sepsis, a life-threatening complication of infection, following a miscarriage. Her miraculous recovery in 1942 signaled a new era of antibiotic therapy that was to revolutionize medicine. Fleming, after receiving the Nobel Prize in 1945 for his discovery along with Florey and Chain, warned that bacteria could become resistant to penicillin if too little or too much were used. But his warning would go unheeded in the excitement over the first "broad spectrum" antibiotic that could successfully treat many infections before they were even diagnosed. Worse, the eventual large-scale use of antibiotics as growth enhancers in livestock, a vastly bigger commercial market than medical use, was a factor that ultimately triggered widespread antimicrobial resistance. Moving through the food chain from livestock to farm workers, meat, and then consumers, antibiotic-resistant bacteria shared their resistance genes with other bacteria everywhere along the way. Today, it is common knowledge that many bacteria have become resistant to penicillin and many other antibiotics. These superbugs have become a menace, especially in hospital settings, where the vulnerable patient population makes a fertile ground for bacteria. I could only hope that if it was an infection that had led to Tom's pancreatitis, that this third generation of a World War II miracle drug would hold it at bay.

••••••••

Dr. Busiri seemed concerned but confident as he continued taking Tom's medical history. Then he glanced around us, leaned in closer to Tom, and spoke in a hushed tone.

"I need to ask you a sensitive question," he said to Tom. "Do you drink alcohol, and if so, how much and how often?" He asked this the way a doctor back home might ask if we shot heroin. Tom rolled his eyes and motioned for me to reply in his stead. I was getting used to it.

"Yes," I said, a little too cheerfully perhaps, but it was the first easy question we'd come to. "We drink wine. Often. I can tell you exactly how much he drinks. Each day, we open a bottle, and I let him have one glass." Dr. Busiri turned from Tom to me.

"You *let* him?" He shook his head, as if in disbelief—whether over my role as a woman who gave orders to my husband or a wife who allowed alcohol consumption wasn't clear. Regardless, he clucked disapprovingly. "Well, no more alcohol. The pancreas is very sensitive."

Tom piped up; his voice a mere whisper. "Steff will be elated. All the more for her." His sense of humor was clearly intact, even if it fell flat with everyone but me.

Later that morning, the call came that an appointment had opened up for a CT scan. We piled into the ambulance, squeezing close around Tom on his gurney. The CT clinic occupied a space in what amounted to a small commercial strip mall, and the surrounding hubbub reminded me of similar scenes during our many trips to India. Swarms of people bustled around. A teenager on a bent bicycle swerved around the ambulance.

Several women strode past, carrying pots of vegetables on their heads and children in slings across their hips. Across the street, a man hacked at a goat carcass that hung upside down in the doorway of a halal butcher.

At the doorway to the CT clinic a queue of people waited, pacing and smoking. The line wound around a litter of kittens that suckled their mother's dry teats on the doorstep. The mother cat hissed as she was forced to make room for Tom's gurney, carried inside by six men, with me and Dr. Busiri in tow. Inside, the waiting room was crammed full of wobbly plastic chairs; on every seat was a woman in a black hijab. Several glared at us, and I could understand why. It looked like we were jumping the queue. What they couldn't know is that we had been waiting standby at the clinic for the call to come. I looked at them apologetically as Tom's gurney was pushed into the back room, but there was no way to explain.

No way to explain. That was quickly becoming the defining sense of foreboding that I couldn't shake. There was no way to explain any of this, no way to get a handle on it. What had begun as a seemingly routine run-in with food poisoning was turning out to be something quite different—and disturbing. Everything about this was disorienting. Tom was a supremely hardy guy; he'd been fine—and then suddenly he wasn't. In the barely thirty-six hours since Tom had fallen ill, every step forward—from the ship to the clinic to the diagnosis of pancreatitis—seemed to be a step back, spiraling us only deeper into this medical crisis, not out of it. Every step took us further from any expertise I had that could even remotely help Tom. It felt like we were channeling Apophys, the god of chaos.

6

THE COLONEL FROM AL-SHABAAB

Later that day, as we dozed side by side in the operating room, Tom on his gurney, me on mine, Khalid appeared to tell me that I had a visitor: an officer from the tourist police. What now? Was getting sick in Egypt a crime? He assured me it wasn't, but any incident involving a foreigner needed to be investigated.

I was ushered into a small cubicle in the clinic while Khalid stood quietly in the corner, observing. Officer Aziz sat before me, chain-smoking Camels while absently toying with his bushy black mustache. A portly, middle-aged man, he wore an ordinary khaki uniform in typical military style. At his side, I noted calmly, a gun poked out prominently from his holster, admittedly not something you'd expect in a medical clinic. But right now, the truth was that his inquiry and the power he held to stall our effort to leave felt more threatening than the gun. With a full head of black hair and thick eyebrows, his expression seemed imposing, even menacing. Or maybe I was imagining that. Tom had often described how the mind works to make sense of sensory input—what we see or hear, for instance—but may also filter our perceptions through implicit biases of our own making.

This was central in his decades of work with schizophrenics, but it plays out for any of us in everyday circumstances, shaping our perceptions, our behavior, even our dreams. So I tried to quiet my fears, be calm, and think objectively about our situation.

In broken English, he asked me a series of questions, which included whether or not we had been attacked, robbed, or poisoned by an Egyptian. I assured him we had not. In fact, I stated emphatically, we were being taken care of with the utmost care and respect. Officer Aziz beamed at my remark and sucked hard on his Camel, flicking the long ash onto the floor. Pointing, he asked me to sign an affidavit. It was several pages, the paper was so thin that you could see through it, and it was all in Arabic.

"But I can't read Arabic. Do you have an English version?" I said in the most polite tone I could muster, pushing it back toward him.

"You must sign," he replied, pushing it back.

When I refused again, Khalid crept from the corner and explained my discomfort to the officer. They discussed the matter for several minutes before Khalid finally relented. He picked up the pages and looked them over quickly.

"It is okay," he said. "You should sign." More quietly, he whispered, "They just don't want a bad scene in the paper where tourists are complaining of being attacked or injured." I signed the form and handed it to him, worried that I'd live to regret it, but I was more worried about Tom and whether he'd live at all if we didn't get to a hospital equipped to treat him.

Officer Aziz smiled thinly, folded the document and pocketed it, hoisted his pants and gun belt up, and strode out the door. Khalid walked with me back toward Tom's gurney. He explained that he had been summoned back to Cairo, but would leave me in the capable hands of a local tour company that would help with anything we needed. He went to shake my hand, but I gave him a hug instead. As I watched him walk away, I blinked back tears. Suddenly, I felt terribly alone.

Over the next forty-eight hours, despite getting IV antibiotics, Tom's condition continued to worsen, and I spent hours on the phone with reps from the travel insurance company, basically begging them to get us out of there. After speaking to several of their medical personnel, I learned that for Tom to be medevacked I needed to prove that he required a "higher level of care." This was delicate, since I didn't want to insult the medical staff of the clinic, who were doing the best they could with what they had. But by now, Tom's condition was clearly deteriorating. He could no longer get up and walk to the bathroom, even with help from me and a nurse. The nasogastric tube continued to siphon off copious amounts of greenish-yellow fluid. His breathing had become ragged, and one of the nurses had fitted him with a face mask so he could receive oxygen.

Originally, Dr. Busiri had appeared convinced that his clinic could handle Tom's care, but after the radiologist's reading from the CT finally arrived, Dr. Busiri's confident tone shifted, too.

"There is no obstruction," he said slowly. "However, there is a high probability of complications, which would likely occur within the next twenty-four to forty-eight hours."

"What kinds of complications?" I asked him. But his phone rang and, appearing visibly relieved at the opportunity to exit, he hurried off.

Left to my own thoughts, I tried to shrug off my growing sense of panic and the image of some invisible monster waiting in the wings, growing stronger by the moment as we fiddled with clumsy human efforts to see it revealed in shadowy scans and blood tests. When I'd texted Chip about the pancreatitis diagnosis, he called me right back, and he wasn't as relieved as I'd thought he'd be. Although he was satisfied that the clinic doctors had prescribed an antibiotic regimen he would have prescribed himself, pancreatitis was serious, and he felt certain there was something more going on. He just didn't know what or whether the clinic had the resources to tackle whatever turned out to be the cause. That was worrisome. Chip had been a well-known leader in HIV/AIDS research in the eighties, when I was just starting in the field, and I'd met him first in his published research, which was groundbreaking and, at times, visionary. He'd told me once that he was drawn to the big challenges—that when things got comfortable, he got bored. We were kindred spirits in that regard, and I respected his zeal for tackling medical mysteries, but right now I wanted Tom's case to be blessedly boring.

Unlike Tom, I could eat and needed to. I left the clinic to grab a bite and ship some of our luggage home. The travel insurance agent, Carol, had told me that on a medevac flight, we would be allowed only one duffel bag each on the small plane. It's surprising how much stuff you realize you don't need when the only carry-on you really care about is the irreplaceable man on the gurney.

Barely through the clinic front door, I could hear Tom bellowing as I approached his room. There, on the floor beside his bed, was a pool of urine, and he was in an uproar.

"Tell them I need a bedpan! They don't understand me!"

I turned to one of the nurses, who was standing there flustered.

"Bedpan?" I said. "For urine..." I motioned to Tom's privates. The nurse waved her hands in dismay.

"La! La! *No! No!*" she uttered, backing away, and then launched into a tirade in Arabic.

With a start, I realized that she thought I was asking her to touch or see Tom's genitals, a major taboo. At the same time, Tom thrust a piece of paper toward me; he had attempted to draw her a picture of a bean-shaped kidney in an effort to explain his request. At any other time, I would have laughed uproariously and so would Tom, but at the moment, this was no laughing matter.

Suddenly, one of the doctors appeared, having been summoned by a nurse. I recognized him as one of the doctors who had arrived to meet us the first night. Dr. Abboud, a thirty-something man, balding, a few wisps of hair swept across his forehead, matching his weary demeanor. The clothes beneath his white coat looked unkempt, like he had been sleeping. I explained to him that Tom had been requesting a bedpan unsuccessfully for the last few hours and as a result, there was now a puddle on the floor.

"I told them to use Google Translate!" he exclaimed, throwing his hands up in disgust.

"It didn't translate properly!" Tom barked back.

Dr. Abboud translated for the nurse, who disappeared and then reappeared with a bedpan. It was the only one in the clinic, but it was hopelessly cracked.

Just then, my cell phone trilled. It was Carol from the travel insurance company calling back. I filled her in on the morning's drama; I could hear her typing furiously on the other end. She asked to speak to the doctor, and he disappeared down the hall with my cell phone. A half hour later he returned and handed the phone back to me. Carol was still on the line.

"The doctor says your husband is psychotic," she told me.

"What? That's ridiculous," I exclaimed, turning my back so Tom couldn't hear. He would be outraged. Tom was clearly disoriented. But psychotic? No.

"Let me speak to him," she ordered. As Tom spoke to Carol, I listened to his side of the conversation, trying to get a sense of what she was trying to find out. Tom was mostly answering yes or no. As he tried to get more comfortable so he could hold the phone more easily, I saw him gasp in pain. "I don't remember," he told Carol in response to a question I couldn't hear. Then he turned to me. "She wants to know when I was last given pain meds."

I looked at the wall clock: ten a.m. "Sometime last night," I told him, and he repeated it back to her. I could hear Carol's voice get louder on his end of the phone. He handed it to me. "She wants to talk to the doctor again."

I went to look for Dr. Abboud and found him in a back room, which was furnished with a small cot that was covered with a blanket and a pillow. He grabbed the phone from me, exasperated, shooed me away, and shut the door in my face. A few minutes later he emerged and almost threw the phone at me in sheer frustration. I retreated back to the operating theater, hoping that my phone battery would last a little longer. Carol was livid, and after venting for a moment shifted into administrative *sotto voce* to declare her verdict: "I have now confirmed that your husband's pain is not being managed sufficiently, and that the medical equipment in the clinic is insufficient for his needs."

Bad news had never sounded so good. This gave her exactly what she needed to justify the airlift, and unfortunately it was true. All there was left to do was wait. Carol's team needed to process the medevac order, get their medical director's permission, identify the hospital that would take Tom—the short list was London, Istanbul, or Frankfurt—and finally arrange for the plane to transport him.

The next hours ticked by so slowly that I thought the clock on the wall was broken. Antibiotics were still being given as a just-in-case measure, much as they are prescribed so routinely back home, to prevent a possible infection. But if the antibiotics had had any effect, it seemed negligible. Typically, if antibiotics are what's needed, they kick in within a few days, but there was no sign of it for Tom. His pain was unremitting, and the murky fluid from his stomach continued to drain through the tube in his nose into the collection bag that hung by the bed. Nobody knew what was causing the inflammation in his pancreas.

At least I heard good news from Chip. He knew the chief of medicine at one of the top hospitals in Frankfurt, and also had colleagues who could help if we ended up in London. As I waited bedside to give Tom the details, I repacked our belongings so they were ready to go. Our duffel bags were still too heavy, so I left the *Vogue* magazine for the young fashion-conscious nurses, along with some toiletries I hoped would find their way to someone who could use them.

The travel insurance company had emailed me several pages that needed to be signed and faxed or emailed back, but there was no fax or scanner in the clinic. It took several hours to find a hotel that could help me with this task. Every obstacle or inconvenience reminded me how resourceful the local people had to be, day in and day out.

Thankfully, Tom slept while I was gone, but his sleep was fitful. And fearful, as I could see from his expression, even as he slept, when

I returned and sat bedside. When he woke, he motioned me closer. His eyes were wide and his pupils dilated.

"They're experimenting on me," he whispered. "That doctor—he tried to blow hookah smoke inside my oxygen mask!" I looked at Tom incredulously. He glared at me. "You don't believe me," he said in an accusatory tone.

He was right, I didn't. But admitting to that would only alienate him further. Perhaps the doctor had been right after all; Tom was becoming psychotic. What would the sane Tom say if he were dealing with one of the schizophrenic patients in his studies?

"Honey," I began, using a tone as calm as I could muster. "It doesn't matter what I believe; it matters what *you* believe. So, if you believe that the doctor is blowing hookah smoke into your oxygen mask, you'll be anxious and won't rest. I'll stay beside you and make sure no one does anything to you, okay? We are getting medevacked in the morning. This is our last night here, so just hang in there with me." That seemed to appease him. I dug out his noise-canceling headphones from the duffel and put them over his ears. He closed his eyes and slept.

Twelve hours after Carol and I had hung up, my phone rang. The details were confirmed. Two doctors would arrive the next morning to determine if Tom was fit to fly, and with any luck, we would be whisked off to a hospital in Frankfurt in a small Lear jet equipped for medical transport. I texted Chip with the update, and Tom's daughters, too.

I'd barely closed my eyes when Tom woke me in a panic.

"Steff, Steff, wake up!" He looked anxiously around the room to see if anyone else was with us in the operating theater, but we were alone. "The colonel is coming! They're going to kill me! We have to get out of here!"

"What the hell are you talking about? What colonel?" Was this paranoia? Or was he hallucinating?

"The colonel from Al-Shabaab!" Tom screamed.

All right then: there was no doubt about it. My husband was now psychotic. What should I do? Over the next half hour, we spoke in hushed tones to one another. Tom was getting more and more agitated. I could tell he was feverish; sometimes he asked that the air-conditioning be put on, other times he was shivering. He insisted that while I was sleeping, one of the nurses had come to warn him that the colonel was coming and that he should run.

"But honey, the nurses only speak Arabic, remember?" I pointed out to him quietly, praying that this would convince him to give up this fantastic claim. Tom blinked and lay on his back, looking at the ceiling, thinking.

"OK. You're right," he finally said. "But I am sure about the hookah pipe. Totally, totally certain." A minute passed and Tom was still looking at the ceiling. "My God, am I losing my mind? I can't even trust myself!" he started to whimper. This was another shocker. I had only seen Tom cry once before, when his dad died.

"You can trust me," I told him fiercely. "I will protect you." I realized how preposterous this sounded—nothing seemed further from the truth. I was overwhelmed and about as clueless as I'd ever felt in my life. Like the hissing mother cat on the steps of the CT clinic, my protective instinct was in overdrive, but with no resources to back it up.

Besides, Tom was always the one to protect *me*, instead of the other way around. But hearing me say this calmed him down and he drifted back to sleep. And I was hopeful.

The next morning, Tom was weaker and more feverish. Once in a while he would ask me what time it was and whether the plane had arrived yet, but he seemed to have lost track of whether it was day or night. I watched the clock nervously. Finally, just before noon, the operating theater doors burst open and two women strode in. Both were over six feet tall and wearing Doc Martens boots that further established their commanding presence. They carried black backpacks full of medical equipment. In clipped German accents, they introduced themselves, first names only, as Anneke and Inge, physicians from the medevac company that had been contracted to fly us to Frankfurt. As they took Tom's vitals, they spoke German to one another, while two nurses and Dr. Abboud watched silently.

"When was his blood sugar last taken?" Inge asked Dr. Abboud sternly, in English.

He rifled through the pages of the medical chart and replied, "Last evening."

Inge and Anneke exchanged glances. Anneke was wearing electric blue eyeliner straight out of the 1980s, which made her eyes widen and look even more alarmed. Tom's blood sugar level was seventy-five, which was dangerously low. "His blood sugar should be tested every four hours under these circumstances!" she snapped. "He could end up in a diabetic coma!"

Dr. Abboud blanched. "We do not have enough supplies for such frequent testing," he replied defensively. Anneke turned to him and said sharply, "Well then, perhaps you should not be in the business of running a medical clinic."

I was mortified, embarrassed for Dr. Abboud and yet angry that Tom's care was not being properly managed, as I had begun to suspect. Within minutes, Inge had given Tom dextrose to increase his blood sugar and stabilize him for transport. Anneke started him on a morphine drip and IV Tylenol to bring down his fever, as well as several other mystery medications that dripped into his arm, one by one. Then Inge spoke into her cell phone in German, presumably to the pilot who was waiting at the airport. Anneke turned to us and said the words I had been desperate to hear.

"He is fit to fly. We will depart immediately."

As the doctors scurried to sign the last of the required paperwork, I sat beside Tom and wiped his brow with a cool cloth. He wavered in and out of consciousness but opened his eyes briefly. "Are you seeing what I'm seeing?" he asked.

"What do you see?" I said cautiously.

"Angels."

"Yes," I replied with a smile. "And these angels are rescuing us."

Tom: Interlude I

"This is a part of what you'll need to know to survive," my father says to me. I believe him. He's given me a .22 and taken me to the woods, taught me to shoot, to hunt, and to skin a rabbit for the campfire. We are at our cabin in the mountains built by my mother's grandfather. I know the way by heart—memorized it when I was ten, trained for the day the Russians would drop a nuclear bomb on us. I can hear the air raid siren—or is that an ambulance? My classmates duck and cover, but I know that I am supposed to take off and walk the sixty miles to the cabin where we all will find our way. Find each other. In safety.

In the cocoon of my father's Navy-issue hammock stretched between two ancient oaks, I can smell the rotting leaves, the humus, layer on layer of leaf mold making a soft, spongy carpet around me, like the generations of my family that walked upon them before me.

Family lore has it that my great-grandfather walked the Trail of Tears as a child. At age sixteen, my grandfather ran away from home, where he was being raised by his sharecropper uncle, to fight in World War I. The world can collapse around you, but you can survive, they tell me. Resilience is the sad, strong legacy you carry in your bones, they say. They mean well, but metaphors are not medicine. Where does it reside, exactly, in my DNA? I need to ask Steff. I can feel the world collapsing now, from the inside out.

7

A DEADLY HITCHHIKER

Goethe University Hospital (Universitätsklinikum), Frankfurt
December 3–4, 2015

Six hours and three thousand miles later, with a brief stop to refuel the small plane in Bosnia, we taxied to a stop on a military airstrip in Frankfurt. It was night, the far end of a travel day that Tom had spent in a state of suspended animation, drugged for pain and stabilized for transport. A waiting ambulance skirted through evening traffic to the sprawling urban campus of the Goethe University Hospital, known to locals as the Uniklinik.

A massive research hospital and premier medical center, the original buildings of the hospital had been built in 1914. If the Luxor clinic in Egypt was emblematic of the challenge of introducing modern medicine in resource-poor settings, the Uniklinik was a monument to centuries of investment and advancement in Western medical science. It was a sleek, contemporary building of glass, steel, and concrete that exuded the efficiency of German engineering. Inge and Anneke shook my hand officially and bid us farewell.

Tom looked more rested now, ensconced in a nest of high-tech monitors and IVs, but his face was still pallid and drenched in sweat. The monitors tracked what could only be considered a fragile peace as vitals go, and IV bags of fluids, antibiotics, and pain medication hung like holiday tinsel from a rolling rack. He drifted in and out of delirium as the nursing staff came and went. Each time someone entered, they donned protective gloves and a plastic hospital gown that they discarded before leaving. They continually monitored his temperature, heart rate, oxygen level, and blood pressure. From the digital thermometer reading of 102.5 degrees Fahrenheit, I knew he had a fever, but the other readings were a mystery. A new line was inserted for IV fluids and antibiotics, and the collection bag that filled with bile from the tube in his nose was swapped for a bigger one.

Tom's eyes blinked open and he caught my gaze.

"I have the worst hangover *ever*," he groaned. He was dripping with sweat and asked for more pain meds; his head was pounding and his body ached down to his bones. "And I'm so bloody hungry—" he said, stopping abruptly. I must have come into focus, because he gave me a head-to-toe glance and rendered a verdict.

"You look like hell," he said. It was true. I was a hot mess; still in the rumpled clothes from the last few days, and frankly pretty gamey by this point. I smirked and thought of a retort my mother would give. The pot calling the kettle black.

"You, on the other hand—a page right out of *GQ*?" His usual trim goatee bled into a five-day-old beard, but they'd cleaned him up at least, having popped him into a fresh blue gown with polka dots and matching blue socks. He glanced down at his gown and looked confused.

"When did all this happen?"

He looked around the room, trying to reconcile his last memory of our whereabouts—the spartan operating room in the Luxor clinic—with this gleaming, high-tech hospital room, every square inch packed with sophisticated medical and monitoring equipment, amid a flow of doctors and nurses and aides in familiar Western-style scrubs and uniforms. The muted sound of German voices in the hallway outside was barely audible, but offered a reassuring hum of human proximity.

"Where are we?" he asked, in a tentative tone.

Before I could answer, his dilated pupils shifted suddenly to the wall opposite his bed. His eyes widened, his gaze fixed on the white wall, then moved slowly up to the ceiling, and down again.

"What are you looking at?"

"The hieroglyphics," he said, nodding toward the wall in front of us.

I looked at the wall, then at him. He gazed fixedly at the blank wall.

"Tom. We are in Germany. The Frankfurt Uniklinik, which Chip says has one of the top GI clinics in the world. We've lucked out. You're gonna be okay." I spoke soft and slow, in a tone more appropriate for a kindergartener, which would have offended him if he'd been in his right mind. "They have us in an exam room while they figure out which ward you will be assigned to."

"Then why are there hieroglyphics on the walls?" he asked.

I paused and looked him in the eye, searching for some glimmer of sanity.

"There aren't any, honey."

"No, look," he insisted. "They're faint, but they're there. On the ceiling, too. Here—put your hand on the wall," he said, motioning to an area about eye level for me. This was nuts, but I humored him by holding my hand up to the wall.

"Don't you see them?" he asked, but disappointment had already registered on his face. "I'm losing my mind—again—aren't I?" he said, closing his eyes. This time, I didn't answer.

He wasn't ambulatory, but he insisted on trying to walk to the bathroom three feet away. I braced him in my arms and we limped over together, only to discover the door was locked. Like a marionette dangling from IV lines, Tom wavered. When I buzzed the call button for help, a nurse stuck her head into the room and pointed to a portable commode by the gurney. "No bathroom privileges. Infection control."

I was confused at first, then annoyed, but then understood.

The hospital was taking universal precautions, which were standard infection control procedures. Having just arrived from Egypt, Tom could have acquired a foreign pathogen—an uninvited hitchhiker that could put other patients at risk. It was just a precaution. Or was it? Suddenly, it dawned on me that since Tom had a fever, some kind of infection could be contributing to his pancreatitis. My memory flashed to how the German medevac doctor, Inge, had carefully scrubbed her hands and forearms after spending just a few minutes in the Luxor clinic. A sick feeling of dread pierced the pit of my stomach.

****

Within the hour, they had a room for Tom. And not just any room. A transport team arrived to push his gurney from the ER to a room in the intensive care unit (ICU) on the third floor. Of course, Tom was sick enough to warrant an ER visit, and hospitalization was a relief so he could get the care he needed. But the ICU? Wasn't that overkill? I traipsed along, watching one of the porters enter a code on a keypad outside the ICU's rear entrance. The door lock mechanism buzzed, allowing us to pass through a pair of heavy metal doors that clanged shut behind us. Tom would later tell me that the sound reminded him of a psych unit, which made him panic. At the time, he was so out of it that I mistook his panic for pain, one of a long trail of miscues that left Tom alone with his terrifying fears while the rest of us focused on the vital statistics of blood and guts. Focusing on data was my way of reducing the growing crisis down to a manageable size. Tamping down my sense of alarm, I busied about Tom's room, plumping his pillow and checking his blanket, as if this turn of events didn't scare the bejesus out of me.

Since it was now past ten p.m. and I was exhausted and hungry, I kissed him on the forehead and gathered my duffel to leave.

"You're leaving me alone here?!" Tom yelped. The beeps on his heart monitor quickened, and I could feel my own heart do the same. Alone? In this buzzing hive of humanity? Hardly. But his sense of betrayal and abandonment shone in his eyes.

"Family can't stay overnight here, hon. The rules are, well, strict. But I promise I'll be back in the morning, first thing. We both desperately need some sleep. You'll finally be able to get some here—they've got morphine to make sure of that. And I'm going to find the closest hotel nearby. I'll fill everyone in back home, and with a shower and decent night's sleep, I'll be better company when I see you in the morning."

I felt guilty but I was secretly relieved. The closest hotel was a brisk walk, about a half mile away. The duffel was packed for Egypt's warm desert climate, not for Germany's winter, and I was freezing cold even though I was wearing all of the clothes I had with me, in rumpled layers. The hotel lobby was decked out with a twenty-foot Christmas tree and a gingerbread house that was large enough for several small children to play inside while their parents stood idly by chitchatting. I had almost forgotten that Christmas was just a couple of weeks away. A group of carolers in traditional German costume sang "Stille Nacht." I listened to their voices echo through the vast foyer as I waited for the elevator. Silent Night? No way, José. I was in no mood for the festivities; there was nothing to celebrate.

Soon after my arrival, a bellman delivered a package to my room, a shipping box with my name on it. It was a care package from my UCSD staff. Pajamas, a jacket, scarf, socks, and a basket of wine and cheese. My eyes filled with tears, I was so touched by their gesture. I had a long, hot shower, pulled on the PJs, ordered room service, gulped a glass of wine, and collapsed into bed for my first full night's sleep in more days than I could remember.

••••••••

The next morning, I made my way to the hospital early, hoping to catch Tom's medical team as they made their rounds and discussed next steps. Chip's colleague had arranged for me to enter the ICU outside of normal visiting hours, which were four to six p.m. daily. I was there at eight a.m. sharp. Once buzzed in, I noticed a sign posted outside the door to Tom's room, giving instructions in German and English that no one was to enter without a protective gown and gloves. They were stacked nearby. I obeyed; we'd clearly crossed the line from the improvisational setting of a community clinic to the exacting environment of a cutting-edge critical care unit. Everything about the ICU had an imposing sense about it. Order. Attention. Expertise. Precaution. Pancreatitis alone is enough to kill you. Tom didn't need to add a case of the flu or some other communicable disease passed along on someone's hands or clothing. He still looked pale and drawn and now had an oxygen mask on his face, but I was relieved to see that he was in less pain. No sooner had I arrived than his nurse, a cheery fellow named Roy, stopped in and informed us that Tom was scheduled for a CT scan. His Filipino accent was familiar to me, and we shared the experience of far-flung family. Shortly, the same two porters from the transport team appeared and greeted us with smiles.

The CT was only supposed to take a few minutes. Almost an hour later, I was still pacing the floor of the waiting room. When Tom finally emerged, the porters pushing his gurney were not their cheerful selves. Something was wrong.

We'd barely returned to the ICU when a doctor joined us in Tom's room, greeting us with a warm handshake. Dr. Stefan Zeuzem was tall and trim with perfectly coiffed gray hair, and looked to be about fifty-five years old. This was Chip's colleague; a famous gastroenterologist who clearly held a lot of clout within the hospital. After a few minutes of polite chitchat, he was all business. Tom was groggy but awake.

"Dr. Patterson, you have acute pancreatitis with the complication of a pseudocyst in your abdomen. The cyst is an abscess approximately fifteen centimeters in diameter, which is about the size of a football. Americans love football, yes?" Dr. Zeuzem smiled. I flashed on the ghastly image of a festering cyst the size of an American football, and then I realized that he probably meant a soccer ball. That didn't help—even a small kid-size soccer ball was still huge.

"We suspect that you may have a gallstone, which created the cyst. You also have a good deal of ascites in your abdomen—fluid caused by inflammation. I have asked my top GI doctor to perform an emergency endoscopic procedure to investigate the cause of this problem, and to remove any possible obstruction." He paused, then frowned slightly. "I must warn you that this is a high-risk procedure given your current state, but we believe it absolutely necessary. If you agree, the procedure will take place within the hour."

We agreed. What choice did we have? As unlucky as it was that a gallstone had likely caused this cyst, and that the cyst became infected, and that it flared up on vacation far from home, at least we were lucky to be where they could take care of it. Tom gestured for me to sign the required consent form on his behalf and I did so with a shaky hand.

The GI doctor who would perform the procedure was Dr. Friedrich-Rust, who introduced herself with a brisk but cordial air of efficiency. Trim and precise in appearance as well as manner, she asked me to sign a second consent form in case Tom required a ventilator during the procedure, which I knew was a breathing apparatus used for life support. She tucked the paper into a clipboard and disappeared down the hall to the surgical unit, leaving us in the waiting room with two porters.

I was scared to death. This was clearly a do-or-die moment. Tom's condition was now undeniably life-threatening, and in minutes they were going to wheel him away. Would I ever see him again? He was still delirious, but aware enough to be anxious. We both needed a distraction. He lay with his eyes closed and asked me to describe our surroundings.

"Well, it is your classic German minimalist architecture. Lots of steel, granite, and stark colors. Except, of course, for the splashes of color on our gowns." Tom opened his eyes.

"Your gowns?"

His incredulous expression suggested that he thought I was talking about ballroom dancing, not hospital gowns. The porters glanced at me quizzically.

"Yes, Marta over here is wearing a beautiful green gown, the color of spring. Me, I am wearing yellow, like the sun." The second porter, Paulina, piped up cheerfully, "And I am wearing blue like the sky."

"So, it's almost like the Four Seasons or something," I told Tom with a wan smile. Then it was time. I gave him a quick kiss, and just like that, he was wheeled away. He gave me a little wave, and was gone.

The procedure took over an hour, and when Dr. Friedrich-Rust was done, she joined me in the waiting room and plopped down beside me, removing her face mask and surgical cap.

"The procedure was a success," she said, with calm reassurance. "He did not require the ventilator. I removed a small gallstone of four millimeters in size from his common bile duct; there was at least one smaller gallstone present as well. I cleaned up some necrotic tissue—dead tissue—in the pancreas. I also placed two pigtail stents in the pseudocyst so that its contents now drain into the stomach, so it will hopefully shrink. I wish I could have removed more necrotic tissue, but your husband's breathing was very difficult, and he could not withstand more sedation." As she disappeared down the hall, I was still running her medical update through my brain's version of Google Translate.

Back in the ICU, Tom was a little wonky from the sedation, but he was talking.

"So when do we get to go to our room?" he asked me expectantly, as if we were in the hotel lounge waiting to be assigned our room at a *real* Four Seasons.

"Darling, *you* are *in* your room."

Tom craned his neck and inspected his hospital room as if for the first time, including the huge dispenser stack of medicines attached to a contraption with giant arms that monitored his vitals. He shook his head disapprovingly and clucked. "This place could use a *serious* upgrade," he said. "Go and tell TripAdvisor."

•••••••••

A few hours later, Dr. Zeuzem stopped in. He had asked Dr. Friedrich-Rust to take a sample of the fluid inside the pseudocyst, which they expected to be clear if it had only formed recently. He held up a flask that contained a murky brown fluid. The pseudocyst size and contents suggested that it had been there for at least a month. I was stunned. It didn't take a rocket scientist to understand that this was not a good sign. The pseudocyst was "super-infected" with at least one kind of microbe, he told us. They couldn't be sure what it was until the lab results were back in a few days.

I was still trying to wrap my head around the idea that this pseudocyst had been there even *before* the trip, for *weeks* before, and that it was now the size of a football, with some kind of microbe lurking in it. I knew that the doctors needed to know what they were dealing with to decide how to treat it, but there were no rapid tests readily available to identify which pathogen was the culprit. Meanwhile, they'd placed the stents in the cyst to drain into Tom's stomach, where they hoped the gunk inside could move through the digestive tract so he could poop it out. The plan, the place, and the people were reassuring. But so far, Tom's condition was only getting worse. At home, I was an expert in global health, but now I was getting a crash course in global illness.

I was scared and needed someone to talk to, but who? It was morning back home. I called my parents and they confirmed that they could extend their stay to house-sit. I called my closest girlfriends, Michelle and Heather, who both lived in Vancouver, but got voice mail.

Then I called my son, Cameron, and told him what was going on. A night owl, he had seen my Facebook posts, so he knew that Tom was sick, but he was shocked to hear *how* sick. As we spoke, I was acutely aware that it was almost the anniversary of his dad's death—Steve, my ex, had died of a massive heart attack skiing at Whistler on December 12, 2012. Although we had been divorced for ten years at that point, his death had hit me like a ton of bricks and was even harder on Cameron, who has Asperger's syndrome. He'd struggled with depression for over a year, but, now twenty-three, he was living on his own and had recently landed a job with a company that hires and trains people with autism spectrum disorder to be software testers. He had just finished the training and was looking forward to starting his first contract. Neither Cameron nor I tend to easily express our feelings—at times we can seem so focused on practical details that we may seem indifferent to the emotional dimension of the moment. But in this moment, I was struck by his empathy and supportive response. He said that he understood and told me he loved me. I wanted to tell him we'd celebrate his birthday soon, but I couldn't bring myself to imagine two weeks beyond the moment. I always knew that when I married a man two decades my senior, he was likely to die before me. But Tom was the healthiest person I knew. *Was.*

My cell phone pinged with a text from our friend Davey, who was back in San Diego after his Thanksgiving vacation. Chip had briefed him. *Call anytime.*

An infectious disease doc who did double time as a researcher in the same department as Chip and I, Dr. Davey Smith had become a close friend in recent years, especially to Tom. He had grown up in rural Tennessee, where I can only imagine that coming out as a gay man while enrolled in med school had taken tremendous courage. Davey and Tom came from similar humble beginnings. They'd met at a party at the home of Chip and his wife, Connie, and bonded over stories about eating roadkill, rodents, and strange game. Tom one-upped Davey with his story of eating a capybara when he and a group of fellow students were starving to death in the Colombian jungle. Davey had been duly impressed; the closest he had come was eating a few possums. Tom and I had attended Davey's wedding ceremony to his partner, Asher, a year ago, where we had hammed it up with Mardi Gras costumes in a rented photo booth.

I retreated to the hallway outside the rear entrance to the ICU, where it was always very quiet. Davey answered on the first ring, and we skipped the pleasantries. He peppered me with questions. Was Tom septic? Was he on pressors? What are his vital signs? What antibiotics was he getting? He spoke with his typically gentle voice, but I could sense his urgency, worry, and growing frustration that he could not determine how bad the situation was because I could not answer a single question. Hell, I barely *understood* them. I heard myself giving excuses. They have rounds early morning, before I'm allowed in, so how can I ask the doctors anything? All of the bags on the IV pole have labels in German, so how am I supposed to know what meds he was getting? And what the heck is a pressor? My voice sounded frantic, as if it were someone else's. The more rapid and shrill I sounded, the more slowly and gently Davey spoke. I had seen him use this technique with a student of ours who had failed his comprehensive exams. I was starting to lose it. *Suck it up, Princess*, said the voice inside my head.

It was my right to ask questions and get answers, Davey reminded me. I could ask to speak to the charge nurse, or the attending physician. I could ask for a printout of his lab values and learn what they meant. I could take a cell phone shot of the monitors and text it to him and Chip so that they could weigh in.

As Davey's words started to sink in, my brain fog began to lift. But Davey wasn't finished.

"Steff, the only reason Tom is still alive is because you got him mede-vacked out of Egypt. *You* did that. So trust your gut. Get informed. I know you are terribly stressed out and tired. Tom is in great hands, but he needs you to be his advocate. That's your job now. That's what *anyone* in his situation needs. Do you think you can handle that?"

It was exactly what I needed to hear. Davey was chiding me in a gentle way that only Davey could. Tom had been rescued, and I had been passively waiting for someone to rescue *me*. No one was going to do that. I'd have to do it myself.

••••••••••

How could something as small as a four-millimeter gallstone create such havoc? After Davey's pep talk, it was high time to find out. I started by asking Roy, the nurse, how many pressors Tom was receiving. Three. That was "high-pressure support" to keep his blood pressure up, according to Roy, who also brought me Tom's labs. Although the report was in German, I used Google Translate to figure out most of the lab markers when they weren't obvious. The lab values that were abnormal were asterisked; there were so many, the page was covered in snowflakes.

While Tom slept, I accessed the hospital's guest internet connection and read up on gallstones, pancreatitis, and their prognosis and complications. It was a slog. Why had I taken American poetry instead of physiology as an undergrad? Thankfully, Chip emailed me a PowerPoint file used to train med students. It showed the typical anatomy of the human biliary tree—the system that makes, stores, and secretes bile. Simply put, bile is an essential digestive fluid that breaks down fats so they can be absorbed by the body. When something impedes that flow, the imbalance triggers a cascade of complications. The PowerPoint lesson mapped out "the ampulla of Vater," which had something to do with blood supply to the gut, but which, in the moment, only reminded me of Darth Vader, the Star Wars archvillain.

The connect-the-dots medical picture was taking shape, however. It turns out that gallstones are not really stones. They are solid bits of material in the gallbladder that are made up of cholesterol or sometimes bilirubin, which comes from bile. People who have a family history of gallstones, are overweight, or have high cholesterol are at risk of getting gallstones. And if you're going to get them, you want them to be big or small, not in between. That's because the bile duct is five millimeters in diameter, so if you have a stone around that size, like Tom did, it can travel outside the gallbladder and get stuck. That causes a backup of bile, which usually causes pain, inflammation, and sometimes pancreatitis. If the pressure from the fluid builds, it can form a pseudocyst, so named because it is a cyst-like sac, an organic holding tank of pancreatic debris. But unlike a true cyst, the walls of a pseudocyst aren't made of the same specialized cells. Instead, the sac is contained by fibrous tissue or granulated gunk. For all practical purposes, though, there is nothing "pseudo" about it. The danger is real. Whatever it was that swam in the sludge in Tom's pseudocyst, clearly they were hoping to keep it contained and drained. The stents were the best hope at this point, but there was no guarantee this strategy would work.

I spied Dr. Jörg Bojunga, the attending physician, and flagged him down as he passed by. He waited while I degowned and dutifully washed my hands before stepping into the hall. He was tall and wiry, with a white coat that hung a tad too short on his lanky frame.

"Now that the GI doc removed Tom's gallstone," I asked him, "does that mean he is on the mend?"

He shook his head. "That could have been the case if we had caught your husband's biliary blockage early, but now that the pseudocyst has formed, grown so large, and created so much inflammation, it could take months to resolve, and that's if we are lucky."

Months?! I was stupefied, and what did he mean by "if we were lucky"? Enough with my head in the sand. I had to know. "I read that the mortality rate from gallstone pancreatitis is about 50 percent."

Dr. Bojunga looked uncomfortable but answered truthfully. "We are hoping Dr. Patterson will recover from septic shock. He also has metabolic acidosis, due to an excess of carbon dioxide in his blood, which is why he is receiving oxygen. As for the mortality rate—it is much higher than that, I'm afraid." He looked at me sadly, with the compassion that any ICU doctor must summon on a daily basis. "Your husband's condition is grave. With these sorts of complications, mortality is *at least* 80 percent. Likely much higher."

Much higher. The words reverberated in my brain like a pinball machine.

8

"THE WORST BACTERIA ON THE PLANET"

Goethe University Hospital, Frankfurt
December 5–11, 2015

With Tom sleeping soundly, I left him a note while I went to eat and gather my thoughts. The crisp air burned my lungs during my walk back to the hotel. It was below freezing, but I was already numb. I watched other pedestrians scurry by, carrying brightly colored presents from the nearby Christmas market.

I wanted to rewind our lives to the Last Supper and skip it. Or go back to a month earlier when we might have paid attention to some early sign of a gallstone and done something about it. Or to the night at home with my folks before we left, when we'd laughed at my mother's superstitious warning, and not go. I just wanted to know I'd laugh with Tom again.

My spirit sagging under the weight of scientific and clinical data, I decided it was time to message my go-to-guy for all things holistic—Robert Lindsy Milne. An uncanny "empath" who feels he can sense someone's physical and emotional state from a distance, he'd been something of a life coach for the past twenty years. In times of crisis I'd found that Robert's intuitive sense often helped me sort things out and find my way. I didn't know any scientists who consulted a psychic-intuitive counselor. I couldn't find any data to support that it worked, but maybe the science would bear it out someday. And let's face it, I was desperate.

I messaged Robert on Facebook, and he responded instantly via Skype video from his home office in Toronto.

"What took you so long?" he asked, a little indignantly. I don't know how old Robert is, but he has looked a spry forty-five to me forever. Just the sight of his face boosted my spirits. I brought him up to speed on Tom's status and the sobering prognosis I'd just heard. Robert thought a minute, and then rubbed his hands across his face before responding. "I don't think this is his time. But he is very weak, and he needs energy, more than you alone can give him."

He moved closer to the computer screen and adjusted it so he could look me squarely in the eye. "I would call his daughters immediately and tell them that their father needs them. They should come right away. Just knowing they are coming will give Tom a big energy boost, and he needs all he can get right now."

I paced the floor with a glass of wine from the gift basket. "But I'm not their *mother*, Robert. How can I tell them what to do? And besides, they're adults." For the eleven years of our marriage, I'd walked a fine line as the girls' stepmother. As a family, we'd worked on that whole "blended extended" idea. I'd developed a respectful rapport with their mom, Tom's ex-wife, Suzi, and the girls had their own relationship with each of us. When they were younger, we had all vacationed together so the girls wouldn't have to choose time with one parent over the other. Through the years we'd provided parental backup for one another in times of crisis. Now that the girls were young women with lives of their own, I was keenly aware of respecting the boundaries as a motherly adviser. I'd even managed to take on the role of "stepmother of the bride" just a few months before, when Carly and her husband, Danny, had gotten married.

So far, I hadn't screwed it up (at least not hugely), and I didn't want to start now.

Robert was adamant. "They are waiting for the signal from you. Trust me on this. And one other thing—take care of yourself. You've been managing this as if it's a sprint, but it's not. It's a marathon. And when you are running a marathon, you conserve energy. Pace yourself."

My cell phone rang—Chip—and I logged off the Skype call with Robert to take the call. I told Chip about my conversation with Dr. Bojunga. Was it really true that Tom's chances of pulling through could be less than 20 percent? He didn't answer me directly.

"I don't want to scare you," Chip said with his soft Alabama drawl. "But his condition is very serious indeed. I'm hoping that the culture they took from the pseudocyst grows a garden-variety microorganism that they can treat. If not, things could get much worse. I've hesitated to suggest this before, but Connie and I both think it's time to call Tom's daughters. Just in case."

Dr. Connie Benson, Chip's wife, was also an infectious disease doc. She had led the largest network of AIDS clinical trials in the world, and when Chip was in doubt of anything, Connie was his go-to.

I hung up the phone, rattled, and took a big gulp of wine. Several of my most trusted friends and advisers—two gifted physicians and the other a gifted intuitive—in very different ways were reaching the same conclusion on the singular point that it was time to bring Tom's closest family to his bedside. Within a few hours, Frances and Carly were booked on the next direct flight from San Francisco to Frankfurt, along with Suzi and Danny.

••••••••

The next morning, I arrived at the Uniklinik by eight a.m. Tom was still on his oxygen mask, but he was breathing heavily. His voice sounded hollow and raspy, and his eyes roamed the room as if he were seeing ghosts. He faded in and out of consciousness, but in a wakeful moment I squeezed his hand and told him that Carly, Danny, Frances, and Suzi were all on their way and would arrive later in the day. His smile did not reach his eyes.

"Am I going to die?" he whispered.

This was the question I had been dreading. I took a moment to think about what to say. Should I tell him the truth, or sugarcoat it? A few years earlier, Tom and I had sat around the firepit in our backyard, and he told me what it was like when his mother was dying of breast cancer in her mid-fifties. Tom was in his mid-thirties. Although it was clear to everyone else that she only had a few weeks left, Tom's dad, by this time retired from being a motorcycle cop, was in complete denial and refused to talk about it. One day, Tom and his dad dug a hole in the front yard to plant a tree while his mom lay inside the house in bed. Concentrating on shoveling dirt, father and son finally spoke about the inevitable, neither looking the other in the eye. Tom told me that he wished that he and his dad could have spoken more openly so that they could prepare for her death. When she died a few weeks later, their history of avoidance only made her loss harder on both of them. Tom would want me to tell it to him straight.

I stroked Tom's face. "You are fighting for your life," I told him. "Robert says it isn't your time, but if you want to live, you are going to need to give it all you've got."

His eyes closed and he lay so still, it was hard to know if he could stay in the fight.

••••••••

Later, Carly and Danny arrived at the hospital, coming right from the airport. Carly bounded into the room and gave her father a huge hug. Her dreadlocks were long gone, but she still had her long dark hair, which draped around Tom as she laid her head on his chest. Tom smiled and stroked her hair and sighed. Clearly, this had been the right call by Chip and Robert, at the intersection of medical and mystical wisdom.

While Carly and her dad reconnected, I gave Danny an awkward hug. Since he was a musician, I had bought him a pair of Guitar Hero underwear as a gag gift for his Christmas stocking a few years ago, after Carly and he got engaged. I didn't know until recently that in small words next to the guitar, the words *rock hard* also appeared. I hadn't been able to look him in the eye since.

Soon afterward, Frances and her mother, Suzi, arrived. While Suzi waited at the foot of the bed for her turn, Frances approached her dad, her eyes full of tears, and held his hand. Her long brown hair was tied back in a clip, and her face was almost as pale as his. Tom grinned for the first time in days, revealing furry teeth.

"Group hug!" Carly exclaimed, and we huddled together at the foot of Tom's bed. All of us were teary now.

Tom's room was suddenly full of the closest people in his life, but he scarcely noticed. He dozed, in and out of consciousness, mumbling unintelligibly, and sometimes shouted aloud at no one in particular. At one point, he asked when the train was coming and how long we had been standing there waiting. When I pointed out to him that there was no train and he was actually lying down, he opened his eyes and looked around with amazement. Danny astutely picked out why Tom might have drawn this conclusion.

"The cardiac monitor alarms—they sound like train signals," Danny said, pointing to the tower of medical equipment. "Or at least they do to me, and probably to Tom," he suggested. Tom opened his eyes briefly and nodded.

The group of us took turns visiting Tom over the next two days.

It's interesting how low the bar is for "looking better" when the baseline is that someone is lying inert and largely unresponsive in a hospital bed. We'd speak to him, and if we got a hand squeeze or saw his eyebrow move, our hopes would spike that he was coming around. It's ludicrous, in retrospect. But having us all there to share the bedside vigil allowed us to share this small bit of hope and comfort and take turns at catching some shuteye.

Slow improvements showed up in his labs and charts even when they weren't visible to us. And then sometimes in unexpected moments, he would say or do something and suddenly seemed to be firing on all cylinders again. I was lying in bed one morning at the hotel when the phone rang. As I reached it for it, I thought, they're either calling to me he's dead, or—

"Hi, honey!" It was Tom.

"Wow, you're up—how are you feeling?

"Better," Tom said. "Why aren't you here?"

"Um, because it's five thirty in the morning?"

"Well, when you come, can you bring a croissant and juice and a pear and some Coke with you?"

"Sure, but I don't think you're supposed to have Coke—"

"Just bring it, dammit," Tom begged. "I'll just have a little sip."

And just like that, he took a giant step back from death's door, like it had all been just a bad dream.

At the hotel breakfast buffet, I tucked Tom's breakfast requests into my bag, downed a quick bowl of muesli and yogurt, and grabbed my coffee to go for the brisk walk to the Uniklinik. He was gleeful and devoured everything I brought him within minutes, barely stopping to talk. I texted the girls so they could go shopping to get him more.

"Fank oo," he said, giving me a little wave, as he talked with his mouth full.

"Did you say 'fuck you' or 'thank you'?" I asked with a smirk. A half an hour later, neither of us was laughing. Tom was glassy-eyed and in a stupor. His pallor had whitened under a sheen of perspiration.

"Are you okay?" I asked him, hardly believing my own eyes. Suddenly, the look on his face resembled the look Cameron gave me when he was a baby, right before he threw up his bowl of SpaghettiOs all over my nightie. "Oh God, where's a bucket!?"

No bucket would have been big enough. Tom threw up projectile vomit halfway across the room. It was black and viscous matter like nothing I'd ever seen. And it was everywhere. I rang the call button over and over, but no one was coming. "Help!" I cried down the hall. His nurse, Birgit, came running. Her eyes widened as she took in the scene while grabbing her gloves and flinging the protective gown over her head. It took Birgit and an aide over thirty minutes to clean up the mess. Tom watched, dazed, but his color looked much better.

I tried to shake off the morning's crisis with a bit of humor. "Just think, honey," I told him. "If your head had rotated around a few times, you would have been perfectly cast for a remake of *The Exorcist*." Tom looked at me, but did not crack a smile. Lights on; no one home.

Once the room was cleaned up and Tom's hospital gown changed, Birgit and I spent some time trying to clean black vomit from Tom's goatee. It was slow going, and he started to whine.

"How about we shave it off?" I suggested. "I've never seen you without a beard. It might even be fun." Tom shrugged. Birgit was happy to comply, and within twenty minutes, Tom was barefaced. He'd always been kind of movie-star handsome, his shock of silver hair and a goatee giving him the dashing look of a forever-fifty-something. I barely recognized the man in front of me. He looked like his father. What had I done?

In retrospect, the *Exorcist* scene and barbering mishap were just a warmup for worse things to come.

••••••••

Later that morning, Dr. Zeuzem knocked gently on the door and entered. He was gloved and gowned but this time also wore a face mask, so only his eyes showed, showing less of his face than the hijabs the nurses wore in Luxor.

"I regret to tell you that our microbiology lab has now cultured the sample from the pseudocyst that was collected during the procedure the other day," Dr. Zeuzem told us. "The pseudocyst is infected with the worst bacteria on the planet. *Acinetobacter baumannii*. This microbe has been responsible for the closure of several ICUs across Europe in recent months. It is the worst news we could have had."

"*Acineto* what?" I asked, interrupting him. My degree in microbiology was rusty. I was drawing a blank.

"*Acinetobacter baumannii*," Dr. Zeuzem repeated, pronouncing it more slowly: *ass-eh-NI-to-bacter bow-MAHNI*. He wrote the genus and species down for me on the back side of the printout of daily lab values. As soon as I saw the words written down, they rang a bell. I remembered plating this organism on Petri dishes back in my microbiology class in the 1980s. But it required no special handling at the time. Strange.

The rise of *A. baumannii* as a drug-resistant strain had followed a similar path to that of others, like MRSA, that were showing up more and more in the news. Some years ago, *A. baumannii* was just another of the billions of ubiquitous bacteria that coexist with humans in our guts, on our skin, and in soil and water. It generally threatened only people whose immune systems were severely compromised, and even then, it was sensitive to antibiotics and thus treatable. Then it became a resident in hospital settings, where vulnerable populations, indiscriminate use of antibiotics, and poor infection control measures provided the ideal breeding ground for multi-drug resistance.

President Obama had prioritized the ESKAPE pathogens, the seven most dangerous superbugs, for research funding in an executive order; each letter in ESKAPE represented a different bacteria that had acquired multidrug resistance. *S* was for MRSA, the *Staph* superbug we had acquired in Goa. *A. baumannii*, the *A* in that lineup, was the one taking Tom down now. A few months later, it would achieve the number one ranking on the World Health Organization's list of the world's twelve most deadly superbugs. The dirty dozen.

Dr. Zeuzem continued. "The lab is running the antibiotic sensitivity analysis, which will take another day or two, but I have asked them to do so urgently. I must tell you that this is an ESKAPE pathogen that is renowned for antibiotic resistance. Given that you were in Egypt, it is likely that you acquired this pathogen there, which is worrisome. Egyptian strains tend to be highly resistant."

"The cyst also contains a fungus, *Candida glabrata*, which is not unexpected. While we wait for the sensitivity results we already placed him on our best-guess combination of antibiotics, and we will initiate a fungicide."

Within minutes of Dr. Zeuzem's departure, Tom was fast asleep.

Given his delirium, I didn't think his brain had registered the fear that filled the room, but mine had. His pancreatitis, as bad as that was, had been eclipsed now by the presence of this massive pseudocyst teeming with a type of bacteria that was, more often than not, highly antibiotic-resistant. I'd read in a general way about the rise in multi-drug-resistant bacteria, but mostly as a concern limited to specialized hospital settings or nursing homes where there were numerous high-risk patients with weakened immune systems. That wasn't Tom. Or at least it hadn't *been* Tom before now. Nobody could know where and how he picked up the pathogen, but the pseudocyst in his abdomen gave it a place to settle in and fester.

While Tom slept, I googled *Acinetobacter baumannii* and boned up on its epidemiology and pathology. Our pal *A. baumannii* had been discovered in the soil about one hundred years ago by the Dutch microbiologist and botanist M. W. Beijerinck, now considered one of the founders of microbiology and environmental microbiology.

If it weren't for the fact that it was in the process of trying to kill my husband, I'd have to admire the microbe and its superpowers. This bacterial kleptomaniac collected genes from other bacteria that arm it for resistance to antibiotics. Trading these little disks of DNA, or plasmids, like my son, Cameron, had once traded Pokémon cards, it shuffles, deletes, or reorganizes them to craftily evade the host's immune system. Its other nifty biological tricks include growing a slimy capsule that inhibits the immune response. And it creates biofilms, which are complex Borg-like microbial communities that enable the bacteria to survive in extreme conditions—the ultimate evolutionary advantage. *A. baumannii* thrives on all kinds of surfaces, like countertops and door handles, linens, and the hard-to-reach insides of medical devices. They can even stick to body lice.

A. baumannii is nicknamed Iraqibacter, because more than three thousand wounded American and European soldiers and military contractors were diagnosed with it upon their return from fighting in the Middle East between 2003, when the bacteria was first identified, and

2009, the last time the Department of Defense made statistics public. Those were conservative estimates, limited to just those patients tested for the bacteria. At various points during that period, as many as 20 percent of wounded soldiers in military hospitals carried it. Its early links to Iraq fed rumors that Iraqi insurgents had placed *A. baumannii*–laced dog feces and rotting meat in incendiary devices, so that the shrapnel would not only wound but also contaminate its victims. But it was lax infection control procedures in the US military that were believed to have accelerated its transmission, which unwittingly spread it between hospitals in the Middle East, Europe, and the US. In fact, to propagate itself *A. baumannii* proved adept at manipulating not only the microbial world but an entire healthcare system.

This ticked me off, from both a global health perspective and a personal one. While *A. baumannii* went its way undetected or unreported, my husband was about to become some nameless statistic in the CDC's *Morbidity and Mortality Weekly Report.* I stopped reading, hoping that Tom would luck out and not harbor one of the multi-drug-resistant strains Dr. Zeuzem was worried about.

Luck was not on our side. Late the next day, I asked a nurse for a copy of the antibiotic sensitivity results that had just come back on Tom's bacterial isolate. Although the report was in German, I got the gist. Of a list of fifteen antibiotics, all but three were marked *R*, which meant that Tom's isolate was resistant to them. That left only three antibiotics to which it was partially but not fully sensitive: meropenem, tigecycline, and colistin. Colistin is a "last-resort antibiotic" because the side effects can be gruesome. Developed in World War II, it's not exactly a modern miracle drug. I had never heard of the other two. A quick Google search indicated that these were the big guns in the antibiotic world, real gorilla-cillins.

I typed the terms *pancreatic pseudocyst* and *Acinetobacter baumannii* into PubMed. Might as well cast the widest possible net for any relevant leads to treatments. Up popped a single paper. It described a case report of a patient who had been successfully treated with surgery to remove the pseudocyst, as opposed to abdominal drains to siphon off infected fluid. I downloaded a PDF of the article and emailed it to Chip. It was hard to predict what he'd make of it, but his natural instincts as both a clinical purist and pragmatist meant he considered everything for its potential.

Within the hour, a new IV line was started to pump Tom full of these three antibiotics, and the nurse changed the instructions on the outside of Tom's door. Infection control procedures were now even more strictly enforced. In addition to gloves and a gown, everyone was required to wear face masks and disinfect twice before and after leaving his room.

I needed to find out what this meant for Tom's prognosis. I removed my gloves, gown, and face mask, washed my hands twice vigorously, and proceeded to walk deeper into the ICU toward the nurses' station, where the doctors tended to aggregate. As I approached, I recognized one of the doctors who was part of Tom's care team. She was seated and reviewing paperwork with a pencil tucked over her ear, but when she saw me coming, her face became flushed and a scarlet blotch crept up her neck, like an amoeba.

"What are you doing down here?!" she shouted at me.

Dumbfounded, I tried to explain that I had a few questions about Tom's infection, and added that I had diligently disinfected according to the new infection control procedures.

"Didn't Dr. Zeuzem explain to you that this is the worst bacteria on the planet?" Her shrill voice escalated to a fever pitch. "Please, don't come any closer. Just—just turn around and go back. We have many, many sick patients in this ward who are getting transplants or chemotherapy. If these bacteria spread, it will kill them all!"

••••••••

A. baumannii was shaping up as a devious foe. It had hidden inside the pseudocyst and, by doing so, cloaked itself there behind the pancreatitis we'd thought was the big threat. All the while, the Iraqibacter was multiplying and taking over as we killed off all the friendly bacteria—bacteria that help us stay healthy—with the other antibiotics. We'd been played. We'd set things up perfectly for a hostile takeover. If we had any chance of outsmarting this superbug, we'd have to adapt ourselves and our medical arsenal, strategy, and tactics for this new fight.

Tom: Interlude II

The wind is howling around me in the hazy murk. Swirling around me are torn bits of paper, dead leaves, and dirt. The skin on my face is blown taut against my skull; my mouth is open in a scream no one can hear. I can't even close my eyes. I am being forced to watch what is happening around me, the remnants of my life.

I am holding on to a pole with all my might. The gale is so strong it envelops me, whipping at my thin hospital gown and lifting me up, so I am horizontal. If I let go, I will die. I hear faraway voices that are familiar. One is my mother, telling me not to let go. She is calling out to me, but the wind carries her voice away. I want her to hold me and take away the pain. I pull myself close and hug the pole instead.

The pole suddenly swings ninety degrees under me and I am looking down, where I see flames. My entire being is made up of pain that can no longer be contained inside my body; I am a ball of hot, white light. I am on a spit, spinning on a rotisserie. Skewered. As I spin, lightning bolts of white light drip off me into the flames below, making them glow brighter. I am looking into hell.

To live or die, my decision. Hold tight or let go? I am so tired. Must I decide? There are more voices now. Steff. The girls. Others. We are here. I love you, honey. Please don't go, Pappy. Just hang on. We're going home.

Yes. I want to go home.

PART II

Can't ESKAPE

Gentlemen, it is the microbes who will have the last word.

—*Attributed to Louis Pasteur*

9

HOMECOMING

University of California–San Diego
Thornton Hospital, La Jolla
December 12–13, 2015

Home sweet home. Sort of.

Chip pointed out that it helped to be at a medical center experienced with Iraqibacter. Back in San Diego, with military bases and VA connections, they had that in spades.

"Over here, we have more Iraqibacter than you can shake a stick at," Chip had said. "So let's get him home."

Although the flight home from Frankfurt to San Diego was much longer than the one from Egypt to Germany, the Lear jet air ambulance was even smaller than the one before. It didn't even have a lavatory and had just enough room for Tom, four medevac personnel, and the pilot. No room for me. The choice of this super-compact plane had to do with infection control concerns that had almost scuttled the evacuation entirely. Luckily, Chip was able to reassure the travel insurance company that regular contact precautions were all that were needed, so the arrangements could proceed. I flew back to San Diego on a commercial flight. It was gut-wrenching to leave Tom at the hospital, not knowing whether I would ever see him alive again, but in one of his more lucid moments, he urged me to go on ahead so I could meet him at the UCSD hospital upon his arrival.

I arrived home in Carlsbad in the early evening, and unlocked the door to greet our cat, Sir Isaac Newton. The black *M* that arched between his eyes—a signature of the Maine coon—furrowed as he eyed me reproachfully. I had been gone nearly three weeks, and for the last week he had been taken care of by a house sitter after I had finally persuaded my parents to fly home. They'd already stayed much longer than planned. I popped a few melatonin and collapsed into bed. Newton continued pacing the house, moaning. No, boy, Daddy isn't coming home. Not yet. But soon, hopefully.

Now that we were home, I was confident that, as rotten as Tom felt right now, they'd get him through this and home in time for Christmas.

••••••••

UCSD's Thornton Hospital in La Jolla was only a five-minute drive from my office at the university, but I'd never been there before. As I arrived, the morning sunshine and ocean breeze kissed the day with bright promise. The glass doors opened into a lobby that felt more like a Hollywood hotel than a hospital. A doorman in a charcoal double-breasted suit stood by, and the soaring atrium entry bathed the marbled interior in light. A double row of palm trees lined the path to the elevator. Plush chairs were clustered strategically, some for conversation, others for visitors who sat slumped, trying to catch a few *Z*s.

I followed the signs to the Thornton Intensive Care Unit—the TICU—on the second floor, and hit the buzzer as the instructions outside the door indicated. The double doors swung open automatically to reveal a beehive of activity. The unit was small, with only twelve beds—private patient rooms lining the rectangular space—most of which faced the nurses' station. Several nurses and doctors stood behind the long desk of the nurses' station, heads down, reviewing notes. A few others were talking on the phone. At one end of the hall, a group of doctors and some residents I recognized from our department at the university were huddled around their mobile computer stand, like bees around a flower, conducting rounds. Behind the nurses' station was a whiteboard with the first name and first initial of each patient's last name. "Thomas P" was assigned to Bed 8. As I approached Bed 8 with trepidation, I recognized Davey's familiar voice. He was standing inside the doorway, wearing a yellow gown and blue hospital gloves, conferring with Chip by phone. En route from Mozambique, Chip had stopped in Virginia to visit his daughters before Christmas.

"Steff!" Davey exclaimed. He quickly discarded gown and gloves, washed up and stepped into the hall for a hug. Davey's cherubic face, dimples, and bright blue eyes underscored the fact that he had not yet turned forty-five. But he was such an old soul in the way he navigated these fraught straits. I choked back a few tears as I looked past them at Tom who was lying in bed, deathly white. Sleeping or unconscious.

"How is he?" I asked him quietly.

"Stable," Davey replied carefully, as he pushed the speakerphone button with a gloved finger so Chip could join in.

"Sounds like they took good care of him in Germany and during the flight home," remarked Chip good-naturedly, but I noted that he and Davey avoided making even the slightest prognosis. "What we're discussing now is how to move forward. That paper you sent me the other day hit the nail on the head. There are basically two paths to choose from, each of which has its pros and cons. The first option is surgery to remove the pseudocyst, which has the advantage of getting rid of the problem. The downside is the risk of septic shock, especially when we have so few antibiotics left in our arsenal."

Dr. Sharon Reed, head of the UCSD Health microbiology laboratory, was retesting Tom's bacterial isolate to confirm that it was still partially sensitive to the antibiotics they tested in Germany, Chip continued. "They will also look for synergy between two or more classes of antibiotics—maybe they'll find a combination that has an effect that no one of them has by itself. But that mix-and-match experimentation is going to take longer." Chip apologized as he had to ring off, and Davey took up where he left off.

"The alternative is for IR to insert a drain into the pseudocyst to try to siphon off the infected fluid," said Davey.

"IR?" I asked him.

"Interventional radiology," Davey said. These are the mechanics of the medical field who use image-guided technology to perform minimally invasive surgical procedures, inserting drains, stents, filters, and such in places where the sun doesn't shine. Like Tom's belly.

"Inserting a drain is less likely to cause sepsis, but it doesn't solve the problem," Davey continued. "Drains can also get clogged with tissue and dead cells, and if that happens, he could go septic anyway or the infection can spread."

Neither of these options sounded ideal to me.

"The question is whether he's even strong enough to handle surgery right now or not," Davey explained. "The bottom line is that we don't want his *Acinetobacter* getting out of that pseudocyst into his bloodstream. That would be game over."

Sepsis was the game ender everyone wanted to avoid. Beyond the threat that the *Acinetobacter* itself posed, if the bacteria breached the cyst wall and flooded the bloodstream it would likely trigger this five-alarm response by an overwhelmed immune system. In sepsis, the immune system overreacts to an infection by launching inflammatory responses throughout the body. Then it throws the whole system into reverse with an extreme anti-inflammatory response, which creates the chaos that can trigger organ failure, tissue damage, and a steep drop in blood pressure that can kill you. It doesn't take a superbug to trigger sepsis. Any infection can cause it. Sepsis is so deadly because it can strike so fast and move so swiftly to a systemic shock and organ shutdown. In the US alone each year, more than 1.5 million people get sepsis, and about 250,000 die from it. One in three patients who die in the hospital have sepsis, though it's usually just considered a complication of the medical condition that originally brought them in.

Although I didn't recognize it at the time, Tom had already survived at least two episodes of sepsis, each time brought back from the brink with swift action to right the wild imbalances that the infection had triggered. To do that required spotting the symptoms quickly, Davey told me. I was to watch for fever, intense shivering called rigors, clammy or mottled skin, a spike in the heart rate, shortness of breath, a sharp drop in blood pressure, mental confusion, or sudden changes in urine output.

Suddenly the medical team's obsessive attention to infection control measures and their constant monitoring of *everything*—from blood chemistry to bedsores—took on new meaning. *A. baumannii* was a deadly invader, but Tom's own immune response could be the thing that killed him. It was easy to imagine but impossible to know precisely what had triggered Tom's sepsis—pancreatitis, *A. baumannii*, or other organisms that were present—but the complexity of his illness made the risks of sepsis all the greater. No one wanted to add a full-out, uncontained assault by *A. baumannii* to the mix.

How permeable were the pseudo walls of this pseudocyst? Unlike the plastic anatomical models that we remember from high school biology class, at the cellular level, nothing is so cleanly compartmentalized. Take the simple act of eating: aided by bacteria in the gut, nutrients move from the food we eat, through digestion, into the bloodstream, then into our cells through all manner of chemical and metabolic processes. Molecular activity is constant, a turbulent exchange in which physics meets chemistry, testing boundaries where barriers are inevitably breached.

"And we can't forget about the *Candida* that's in there, either," Davey added, referring to the fungus that had been found in Tom's pseudocyst in Germany. "If *Candida* escapes and enters Tom's bloodstream, it, too, can be fatal."

Tom's no-nonsense nurse, Meghan, interrupted us to summon me to the reception desk. "Dr. Tom Savides is on the line for you," she explained.

Like Chip and I, Tom Savides was a division chief in UCSD's Department of Medicine. Savides was the head of gastrointestinal endoscopy. We were used to seeing each other at faculty meetings, but I had learned over the past few days that he happened to be an expert on pancreatitis and its complications.

"I'm on my way there and will stop by to check on Tom," he told me, after we traded pleasantries. "I've already ordered a new CT and I'm about to review Tom's case with Bryan Clary, chief of surgery, who specializes in high-risk surgeries."

After I hung up and briefed Davey, I could tell he was pleased. "At grand rounds last week, Bryan dared the audience by saying, 'Show me a patient at UCSD that's too sick for surgery.'" Davey chuckled softly. "If Tom needs surgery, that's our man."

What the docs didn't say outright, but I would learn later, was that there was some tension among the specialists over which course of treatment to pursue, in part because, frankly, Tom's condition was so iffy and his prognosis so poor. This wasn't just an excruciating decision for me and our family. It was agonizing for the doctors as well. I would later learn that some thought it more likely that he would die than stage a turnaround, whether treated surgically or with interventional radiology. They didn't want to subject him to surgery with no hope of it working. And hope was fading fast.

For Chip, Davey, and the other infectious disease doctors, any intervention posed a risk, but given Tom's worsening condition, leaving the drains in place when they weren't clearing the infection seemed unacceptable. Tom had once railed when someone quipped that he was a medical guinea pig; now he was, clinically speaking, a hot potato. Maybe the more fitting metaphor was a twisted version of Russian roulette: nobody wanted to be the one who killed him.

For me, the most salient—and sobering—fact was that there was no "right" answer to this dilemma. You could parse the data till the cows came home, and the risk-benefit ratio was a toss-up. It was the devil's choice. I still held on to the naïve belief that the doctors would agree on what to do, and they'd pull Tom through.

Davey told me he would be back later, which gave me time alone with Tom. But there is no such thing as "alone time" in an ICU. Meghan hovered like a surveillance drone, taking Tom's temperature, drawing blood, and entering notes into the computer's electronic medical record. With all the commotion, Tom began to stir.

"Honey, it's me," I whispered to him, holding his hand. Tom opened his eyes groggily and gave a half-smile. "Baby," he said tenderly, and gave my hand a little squeeze. "What's happening?"

"He's on a lot of morphine right now, so he'll be really sleepy," Meghan warned me. She was twenty-something, and about five-foot-two with long dark hair that fell down her back in a low ponytail. She expertly adjusted the IV lines and checked the stacked bank of monitors, which was taller than she was. Although petite, she loomed large over Tom, who seemed to be shrinking in size and presence with each passing day.

I nodded, then turned to Tom. "You're in the Thornton ICU," I told him. "Davey was just here, and we discussed the possible types of treatment to get you better. They haven't made up their minds yet. Tom Savides, the GI chief, and Bryan Clary, the surgery chief, need to weigh in."

"Surgery?" Tom asked, rubbing sleep out of his eye with his other hand.

"Yeah, I know what you're thinking," I told him. "It scares the crap out of me, too."

"Yeah, but maybe I should get it over with so I can get outta here," Tom replied. "Enough of this shit."

"Hey, you just got here!" Meghan teased.

"We started out in Egypt," I told her. "Long story." I turned back to Tom, but he had already fallen asleep again. Meghan gave me a quick rundown of the TICU schedule and rules for visitors. Visiting hours were around the clock. Rounds started anywhere from eight a.m. onward, and family members could attend. Shift change was every twelve hours—7:20 a.m. and 7:20 p.m. Nurses were assigned to one or two TICU patients, depending on their needs. I didn't realize until later that Tom was her only patient that day, which should have told me something about how sick he was.

"You can bring in a few things from home to make him or his room more comfortable," she told me. "But take this home with you," she advised, handing me Tom's wedding ring.

"I doubt we'll need anything," I replied, clutching the ring, which was still warm. "He'll be home by Christmas, right?"

Meghan arched an eyebrow skeptically and pursed her lips. "Don't ask me," she said. "But I wouldn't count on it."

We were about to learn why.

A few hours later, Dr. Tom Savides donned the usual gown and gloves and stepped in to introduce himself to Tom and to say hello to me. About sixty, Savides was medium height with a lean build. This was a guy who ate his daily recommended fiber equivalent. His brown hair was clipped conservatively, and he had friendly eyes behind rimless glasses. Tom was awake but groggy, and when he raised his hand to shake Savides's hand, he got tangled in the IV tube attached to it and took advantage of the opportunity to try to yank it out.

"Whoa, there!" Savides said, and grabbed his hand. He explained that, in addition to the nasogastric tube that continued to siphon out infected fluids from his pseudocyst, the doctors had inserted a feeding tube into his nose, this one bypassing the stomach to the jejunum, a part of the small intestine. "Make sure you avoid the temptation to pull that sucker out, or you could hurt your gut. And from the looks of your CT, you have enough going on in your abdomen."

We'd barely greeted each other before I peppered him with questions. The most pressing one: "Have you decided whether he should have surgery or drains instead?"

Savides cleared his throat. "Well, for now, Bryan Clary and I both think that surgery is too high risk. Sounds like you have a nasty superbug in that pseudocyst of yours. I've dealt with some before, but we generally have more antibiotics to treat them. We'll let the infectious disease folks handle that part. But we talked to IR and they are prepared to insert an external drain into the cyst to help it shrink." If only as a temporary measure, this strategy of détente and containment—the decision to leave the superbug undisturbed and untreated for the time being, while more aggressively draining the pseudocyst—seemed to be the bottom line. The problem was, containment could become difficult and might prove impossible. No one could predict how strong the membrane wall of that cyst was. If it were to rupture, the *Acinetobacter* would spill out and spread instantly.

Davey reappeared at the door of the room and gowned up. "Mind if I join in?" he asked each of us. Savides nodded a welcome.

"How long will it take for me to get better?" Tom asked.

Savides was ready for this question. "We have a general rule of thumb," he replied. "For every week that you are lying in bed, it will take five weeks to recover. So, let's say by the time we get you stabilized here, that it will have been about a month since you got sick in Egypt. That means about five months' recovery time."

"Five *months*?!" Tom and I yelped in unison. I was hoping I had misheard.

"I know it's a shock," Savides went on. "But gallstone pancreatitis—even without complications—is on the same scale as a major car accident. The course of this illness is like a marathon, and you might be at about mile eleven—not quite halfway. With any luck, we can get this infection of yours under control and eventually send you home for outpatient treatment," he replied.

Davey raised his eyebrows but kept quiet. Savides promised that his team would be following Tom's case, but at the moment he had to leave for surgery. He was gone in an instant, and Davey filled the stunned silence with a gentle but sobering confirmation of the typical recovery time.

Davey explained that lying in bed with no exercise causes "deconditioning," a weakening of the muscles that makes it hard to walk. Tom had lost forty pounds in the past twenty-one days. Not exactly the image of a marathoner. Robert, my psychic friend, had warned us in Frankfurt that this would be a marathon and not a sprint. A few years previously, we'd all laughed when Robert had told Tom that that by his sixty-ninth birthday, he would be skinny. "You can get there the easy way or the hard way," Robert had said, admonishing him to take better care of himself. It didn't take a psychic now to see that he was doing it the hard way. Cosmic coincidence or not, we were in for a long haul. Tom was going to need a lot of physical therapy once this was over.

"Will I be home for Christmas, Davey?" Tom's voice was impatient and desperate. He was tired of being sick, tired of the hospital rigmarole.

Christmas was only two weeks away. Davey turned to Tom, and his eyes were bloodshot.

"I don't know. I sure hope so. But I'd be lying to you if I didn't tell you that you are really, really sick. We are doing all we can to get you better."

Reality was starting to sink in through Tom's opiate haze and my own thick fog of wishful thinking. We'd both assumed that somehow, once home, back where the doctors had more experience with *Acinetobacter* and the resources and state-of-the-art medicine to back them up, Tom's recovery would be swift and sure.

The look on Davey's face was unmistakable. He wasn't even sure Tom would make it.

10

SUPERBUGGED

UCSD Thornton Hospital, La Jolla
December 14–23, 2015

I quickly became a fixture at the TICU, spending every morning there like clockwork. I learned that to find out how Tom had fared each night, I needed to speak to whoever was assigned as his night nurse, and to call well before shift change, when they were busiest. So, my daily routine began every morning at five a.m. Quick breakfast and occasional shower, and I was at the hospital in time for rounds. As a UCSD faculty member, I was treated with an unearned degree of professional respect, despite the fact that I was not a medical doctor. Ordinarily, I would never have even thought to be present for these detailed clinical conversations about labs and meds or the meticulous routines of nursing care. But Davey and Chip had coached me to step it up in Frankfurt, and this was no time to step back. So, I listened carefully, then wheedled, whined, cajoled, and advocated like his life depended on it. For all I knew, it did.

Keeping track of who was on Tom's care team was a science unto itself. Unlike Egypt or Germany, where the doctors who presided over Tom were relatively small in number, here there were scores of attending doctors who rotated on and off service in the TICU, in part because it was a teaching hospital. The disciplines caring for Tom included pulmonary and critical care, infectious disease, gastroenterology, interventional radiology, and surgery. The critical care docs rotated every week. The other disciplines rotated every two weeks. And the medical residents typically rotated every month. It was enough to make your head spin. And there was no *ICU for Dummies* pocket guide.

I made a conscious effort to listen carefully at rounds to pick up the lingo so I could understand whether Tom's condition was improving or not. It was tough going. The first few days, I could just follow the big picture. There were so many biomarkers of liver, kidney, and cardiac function. His hemoglobin was measured at least twice a day to determine if he needed a blood transfusion. *Bili* was short for bilirubin, an indicator of gall bladder function. Tom's bili was off the chart, indicating that his gallbladder was not a happy camper. His blood sugar was monitored closely since he was now full-blown diabetic after losing about a third of his pancreas due to the infection. And his fluids were documented carefully as "ins and outs." The ins included IV fluids, nutritional supplements, and blood transfusions. The outs included urine, feces, vomit, and the bilious discharge that continued to be siphoned from his stomach and the gunk from a new external pseudocyst drain that interventional radiology had inserted. At least that delicate procedure was a success, and we could see the results as a cloudy yellowish-brown discharge that drained continuously into a bag. The ooze was the enemy we could see. There was so little in this biological battle that was visible to the naked eye. The lab reports, vitals charts, and monitors were our only window into that war zone.

Soon, the language of illness became my second language. At medical rounds each morning, I could follow the conversation now, even contributing to them from time to time, having memorized most of Tom's lab values. *BP*: blood pressure. *RR* or *res rate* meant respiratory rate. Words like *hemodynamics* and *AKI* (acute kidney injury), not previously part of my vocabulary as an epidemiologist, were becoming familiar. I was no longer reluctant to ask questions—or contribute to the discussion, for that matter. "Welcome to wife-led rounds," the charge nurse, Marilyn, quipped one morning. When Cameron was young, he'd call me "Dr. Mrs. Mommy"—I guess that pretty much summed it up now.

It wasn't that I thought I was indispensable. It was that staying focused on the detail kept me from focusing on the larger, scarier picture. I'd learned my lesson in Frankfurt, where being passive and overwhelmed had added another level of crazy to a situation that was already surreal. Now, I never sat still, never just killed time. If I did, the reality of possibly losing Tom would penetrate my brain, and I didn't want to—couldn't—face it. The closer Tom got to death in the TICU, the more I buzzed around in the hive, looking for things I could do to help. Anything. Mostly, I learned to do tasks that the nurses' aides did. How to brush his teeth with the little sticks that had square sponges on the end—*lollypops*. How to adjust the ventilator hose so it wouldn't make that infuriating dripping sound. How to read the lines on the cardiac monitor that measured oxygen, heart rate, blood pressure. How to tell if his feeding tube was nearing empty and how much urine he was putting out per hour. I watched his pseudocyst drain to see how much purulent liquid was coming out and if the color, cloudiness, or thickness of it had changed. What kind of pressors he was on and what their levels were.

I washed his face, put cream on his limbs, filed his nails, hung cards on the walls, made sure the Pandora station was set to something he liked, and of course, joined rounds. And most days, I would sing or dance to at least one song around his bed. Not that Tom was awake enough to enjoy it. But who knew? Nurses would look down the hall and suppress a smile or laugh. They must have thought I'd lost my mind. *How can that woman dance when her husband is so sick?* I danced for my life. For ours. I was with him almost all the time, but I *missed* him. I missed sharing our lives, talking, laughing. He was either semi-comatose or delirious. It made me think of what caregivers of Alzheimer's patients must go through every day when their loved ones are gone but still there. Tending to the details was something I could do. It gave me courage to soldier on.

For a few days after the procedure to insert the drain into his pseudocyst, Tom was feverish. His resting heart rate averaged over one hundred—high—which was indicative of tachycardia, or being "tachy," meaning there's a problem with the heart's electrical system. His blood pressure took precipitous dips. So did his blood oxygen levels, which the nurses referred to as a "desat," short for desaturation. I made copious notes on scraps of paper to keep track of things. The nurses allowed me to look over their shoulder at his electronic medical record, which included handy dandy ranges for each biomarker. Values that were abnormally high or low were asterisked. I'd grown accustomed to seeing those pages filled with snowflakes since Frankfurt. There were different fields for various kinds of tests that were run in panels, along with his microbiology cultures, which had several exclamation marks beside the description: "heavy growth of coccobacilli and heavy number of white blood cells. Moderate yeast cells." A handwritten note was pinned below a whiteboard in Tom's room. "Dr. Sharon Reed following patient. Please page X30778 in case of emergency."

Many mornings, Tom slept and I absorbed what I could of medicalese by osmosis or by asking the nurses and residents, most of whom were more than happy to oblige. Day after day, the quiet cloister of rooms gave little hint of the struggles underway within. Many of the patients were alone. Some died. Although the doctors and nurses adhered to strict privacy rules about the health status of all the patients, I noticed that several rooms had a sign saying DROPLET PRECAUTIONS taped to their doors.

"Do droplet precautions signify that a patient has TB?" I asked Dr. Eric Scholten, the pulmonary and critical care resident who was on service.

"Not usually," he replied. "These days, it's mostly flu."

"Influenza? Really?" I was amazed. I thought the flu typically hospitalized babies and the elderly. "Don't people get their flu shots?"

"You'd be surprised. We have several patients in here who are in their prime, battling severe cases of the flu. People think it won't happen to them. It happens every winter. And we're in California. There are a lot of anti-vaxxers."

Sometimes under the stress of a flu attack, the immune system grows weak and unable to fight off other disease organisms, or the immune system itself can go haywire attacking a secondary infection, causing sepsis and then organ failure. And sometimes someone's unique vulnerabilities—a weak heart, obesity, a serious injury—made something like the flu that much riskier for them.

From inside Tom's room, I surreptitiously watched one of the few other family members who came on a regular basis stop at the nurses' station. She was a forty-something woman with a black pixie haircut and a pink jogging suit emblazoned with JUICY COUTURE. She wore sunglasses and a cup-shaped face mask as she entered Bed 9. The droplet precaution sign swung back and forth as the sliding door slid shut, and she approached a handsome man with dark wavy hair who was lying immobile on the bed. He was probably ten years older than the woman who I presumed was his wife, and his torso showed a slight paunch. The lower half of his face was almost covered by a ventilator and his skin looked waxy. There were several bags of dark fluid hooked up to his IV pole, which I knew meant that he was very sick. On the wall behind him was a family photo of him and this woman in happier times, with three little girls who all looked to be under five years old. As the woman settled into her chair, clutching her husband's hand and a Kleenex in another, we exchanged glances. *I know what you're going through,* I wanted to tell her. *It hurts so much.*

I felt another pang inside too; it was envy. I wished Tom just had the flu and could swap places with the man in Bed 9.

••••••••

My cell phone pinged with a text from Chip.

You in Tom's room?

Yup.

Be right up.

Tom was still asleep, something he seemed to do more and more of these days. I was happy to see Chip, hoping he could explain the microbiology report to me. But when I saw his face, I could tell right away that something more urgent was on his mind. He was renowned for always smiling or cracking a joke, even in times of crisis. Not today.

"So. The labs are back. They're not what we hoped," Chip said in a clipped tone, after he had donned gloves and gown. His shoulders sagged as he sat down on the stool beside Tom's bed. He was not looking forward to what he had to tell me. Inside, I felt my heart start to race and stomach muscles clench.

"The Micro Lab ran the sensitivity analyses on Tom's isolate based on a sample we took from his drain when IR inserted it," he began. "His *Acinetobacter* is now resistant to the last three antibiotics, including the ones we use as a last resort, meropenem and colistin. We're pretty surprised, since he's only been on those for a few weeks."

I was stunned. Resistant to everything? In a matter of weeks?

"What do we do now?" I asked Chip, feeling numb.

"The lab is still testing combinations of antibiotics to see if there's any synergy. It's going to take more time," he replied. His voice did not sound very hopeful. "In the meantime, we just need to absolutely ensure that the pseudocyst fluid continues to drain, because we can't allow it to spread."

I pointed to Tom's IV pole. Every metal prong was full, including colistin, meropenem, two other big-gun antibiotics, and an anti-fungal. Because they still had him on antibiotics, there was an assumption—on my part anyway—that the drugs were doing *something*.

"If the antibiotics aren't working, then what are those for?"

Chip looked me in the eyes as he stood up to leave. "Those," he said curtly, "are to make us doctors feel better."

Oh, shit. The desolation in his voice trumped any hint of hope I'd held out. I'd had blinders on—medically assisted denial. For just a surreal flash, it all felt like elaborate theater, these trappings of treatment, and all of us dropped onto this stage with roles that none of us ever wanted: the patient, the patient's wife, the desperate doctor.

"But Tom Savides said the other day that with any luck, once Tom's infection is under control he might be able to come home and be treated as an outpatient," I countered. But nothing suggested luck was going Tom's way.

"With all due respect to our colleagues, I don't think they know what we're dealing with here yet," Chip responded, getting ready to leave. He stripped off his gloves and gown, and threw them hard in the trash, frustrated. "Tom can't go home for the foreseeable future, unless, of course, we want to bury him. I, for one, am not about to let that happen."

Neither was I, but resolve alone wasn't going to save Tom. After I was sure that Chip had left, I pulled the brown curtains to Tom's room closed, and laid my head down on his chest. And I cried. Hard.

The next morning, as I walked into the TICU shortly before eight a.m., I froze in front of the nurses' station. The whiteboard behind their desk had no name beside Bed 9. I turned around and looked to see the room where the man with dark wavy hair had been lying just the day before. It was empty.

The man with the flu had died.

11

PUBLIC ENEMY NUMBER ONE: UNDER THE RADAR

The work of an epidemiologist is often to deconstruct disaster, working backward from an epidemic to identify how it happened, so we can prevent the next one. We are no strangers to worst-case scenarios for deadly diseases—AIDS, tuberculosis, cancer, heart disease. But even with that knowledge, the cognitive dissonance that has us, as ordinary humans, logically see one thing and emotionally believe another had continued to skew my expectations toward unjustified optimism.

The reality was that, on a societal scale, health leaders had buried their heads snugly in the sand, ignoring the growing peril of antimicrobial resistance (AMR). As if collective ignorance—or denial—would stave off a pandemic. I asked Chip if Tom's *Acinetobacter* had to be reported to the CDC. After all, the Germans had to report it to their national health agency, and that was before it had become even more antibiotic-resistant. Chip told me that in the US, there were no reporting requirements for *A. baumannii*. National hospital reporting for MRSA was just now coming into effect, lagging well behind awareness of the threat it posed. Frustrated, I tried to google information on how many multi-drug-resistant *A. baumannii* cases occurred in the US each year and found *nada*, except for the CDC outbreak investigation on the military cases I had read about earlier.

I didn't get it. *A. baumannii* was getting a reputation as one of the most formidable antibiotic-resistant pathogens. So, why didn't this bug—and others like it—have reporting requirements, too? Without a superbug surveillance system, we had no idea when a new resistance gene emerged or how quickly it spread. We couldn't track who had it, how they got it, or worse, learn lessons from patients that been treated. We were *allowing A. baumannii* to maintain its invisibility under the radar. And it happily did so, spreading quietly. Unreported. Undetected. And now, untreatable.

The scope of the AMR problem had crept up on me, even as an epidemiologist. Several high-level reports that were intended to mobilize a meaningful public health response had somehow failed to do so. There were still so many unknowns. How many people were infected or killed by superbug infections every year? Globally, the most recent estimate was 700,000, but a few years later, that number would be upped to 1.5 million. And even in most affluent countries with sophisticated medical tracking, no one really knew. Reporting was uneven at best.

I had been deluding myself, too, with naïve assumptions, like the idea that modern medicine could lick most of the common things that walk in the door: flu, food poisoning, infections you pick up on vacation. Science and statistics aside, our personal experience of Tom's repeated brushes with parasites and other pathogens over the years had conditioned me to expect medicine—and Tom—to prevail. Even as a scientist who knew how dicey the microbial landscape could be, I generally considered bacteria to be inferior to viruses, less threatening. Apart from *Mycobacterium tuberculosis*, the bacterium that causes TB, most bacteria I dealt with had been fairly harmless. I'd streaked them on my Petri dish in my college microbiology class in the 1980s, no biohazard flow hood or protective gear required. I bought probiotic yogurt, understanding the need to promote the growth of friendly bacteria in our gut microbiome. And when they acted up? Antibiotics could handle it—just as Tom and I always had with our travel dose of Cipro.

We've put people on the moon and developed technology that can take your gall bladder out through your mouth. How could the best doctors and the best medicine at one of the top medical centers in the world be coming up short? How could we possibly be helpless against a species of bacteria that was once so mundane and harmless—wimpy, as Davey had said once. *How the hell did we get here?*

••••••••

Thought to be a companion to humankind since our earliest forebears roamed the earth, *A. baumannii* and many other bacteria for the most part evolved with us—and within us—as relatively benign microbial hitchhikers. At every step, the goal of bacteria has always been straightforward: to go forth and multiply, evolving to match any challenges to survival. These included our migrations across continents and seas and the resulting changes in the flora and fauna of our habitats, as well as in the blood-and-guts microbiomes of our biological selves.

As our civilizations evolved, so did our microbial companions. Crowded living conditions, poor sanitation, air travel, and wars opened new opportunities for bacteria to breed, thrive, and spread relatively undeterred by medical science until Fleming's discovery of penicillin in the early twentieth century. In short order (not even a blink in evolutionary time), scientists developed techniques to make and manufacture antibiotics for the masses. Suddenly pharmaceutical companies were cranking out drugs that could kill bacteria better than anything in human history. Bacteria had to pick up the pace of adaptations to match this new threat to their survival. And *hoo boy*, they did.

Nature equipped them brilliantly for the task: to detect threats, adapt quickly to defend themselves, and pass along their genetic playbook to their progeny and other bacteria. We measure human evolution in millions of years; bacteria do it in minutes. Antibiotic resistance can spread via two routes: reproduction and ordinary contact. As bacteria multiply through binary fission—one cell dividing into two new ones—the new generations of bacteria carry the mutations forward in a process called vertical transmission. Bacteria can also acquire resistance genes through horizontal transmission, as they encounter other bacteria in the air, or those that travel with us on objects or in environments we share.

Perhaps the scariest version of this community gene-sharing involves the plasmids—those Pokémon disks of DNA that can contain multiple genes for antibiotic resistance. *A. baumannii* is especially adept at picking up plasmids from other common bacteria, like *E. coli* or *Staph*, continually expanding its collection of antibiotic resistance genes that it passes on to its progeny.

Resistance genes confer what amounts to molecular chemical weaponry, arising from spontaneous genetic mutations that block or disable the threat. Bacteria can produce enzymes to deactivate an antibiotic. They can reroute an incoming antibiotic and dump it outside their cell wall. They can modify their own architecture to close off access points—receptors—that antibiotics use to attack them. If an antibiotic is designed to disrupt the bacteria's cell wall, any protective barriers, or its reproductive cycle, the bacteria can reconfigure itself to fend off the attack. Some bacteria can even "hibernate" to avoid a predator. They share defensive intel with other bacteria through the use of electrical and chemical messaging, called quorum sensing. Under attack, bacteria that have developed the right mutations repel the attack; those that haven't are done in.

Unaware of the invisible threat of AMR, people pass antibiotic-resistant bacteria along simply by coughing, hand-to-hand contact, or touching surfaces where the bacteria have set up shop—like Pokémon Go, but with the unsuspecting gamers catching superbugs. That's also how overuse of antibiotics in livestock cultivates resistant bacteria, contaminating the animals, the environment, and the food products that people handle and eat.

Concern about AMR was long dismissed as alarmist, despite Fleming himself having warned of it from the beginning. The first sign of trouble came in 1940, when a strain of *E. coli* was found to deactivate penicillin by producing an enzyme that destroyed it. Within two years, several strains of *Staphylococcus aureus* showed penicillin resistance in hospitalized patients. But labs were discovering new and stronger antibiotics far faster than bacteria were developing resistance, at least at first.

We call it Big Pharma now, a $446 billion enterprise in the United States alone, but the industry as we know it was barely advanced from the age of snake oil potions and pills at the turn of the twentieth century. Antibiotics became a large part of what put the "big" in Big Pharma. And with war on the horizon, there was big money to be made in manufacturing these new wonder drugs. Anne Miller's case in March 1942 was a watershed moment in the commercial history of antibiotics. Penicillin was still not in large-scale production and was such a small-batch drug that the 5.5 grams used to treat her—roughly a rounded teaspoon—amounted to half of the US supply of the drug. Within a year, new manufacturing technology was developed that allowed for mass production. By 1945, penicillin was in widespread use, prompting Fleming to sound the alarm again.

"The thoughtless person playing with penicillin is morally responsible for the death of the man who finally succumbs to infection with the penicillin-resistant organism," he wrote, after he and his colleagues received the Nobel Prize that year. It fell on deaf ears among industry, physicians, patients, and policymakers who were not keen to cut back.

Between 1940 and 1962 alone, pharmaceutical companies introduced more than twenty new classes of antibiotics to an eager market. In practice, that meant the release of hundreds of new drugs, because each class contains many related subtypes, each one tweaked in some way that tricks the bacteria, at least for a while. It was easy to believe that there was simply no end to the ingenuity of pharmaceutical research and development. New antibiotics soon followed: streptomycin, chloramphenicol, the family of tetracylines, erythromycin, and the cephalosporins. *So what if bacteria were developing resistance?* the thinking went. *We'll stay a step ahead by continuing to develop new drugs.*

So they said. They were wrong.

More than 150 antibiotics have been developed since the discovery of penicillin, and for the majority of antibiotics available, resistance has emerged and gone global. At the rate that scientists now know that bacteria develop resistance, researchers would need to create about thirty-five new classes of antibiotics each century to stay ahead of bacterial pathogens. Instead, no new antibiotic drug class has entered the market since 1980, and no new class of antibiotics has been discovered to treat Gram-negative bacteria like *A. baumannii* since 1962, before I was born. All bacteria have an inner cell membrane, but one reason that Gram-negatives tend to be more resistant to antibiotics is that they have a hardy outer membrane as well, which makes it tough for many antibiotics to penetrate. In 2015, there was only one novel drug in clinical development that could potentially fight Gram-negative bacteria.

These days, most Big Pharma companies have shut down their antibiotic research labs or laid off researchers. It's much easier (and cheaper) to develop new versions of drugs within known antibiotic classes than to push for brand-new ones. The once massive engines of government funding and private profit that powered new drug development in the mid-twentieth century can now barely keep the lights on. Global health experts have long called for antibiotic stewardship, which includes using antibiotics sparingly and saving some as a last resort. But this only created even more of a disincentive for the pharma industry. One expert compared pharma's reticence for investing in new antibiotics to the public's disinclination to buy fire extinguishers. Why pay so much money for something you might never use?

Bacteria, meanwhile, have continued to evolve rapidly. As bacteria have grown more resistant, the antibiotics we've relied on to treat *Salmonella*, *Campylobacter*, *E. coli*, and other relatively common foodborne pathogens have become less effective. The CDC estimates that one in five antibiotic-resistant infections in humans originate from food and animals. That's because in many countries, animals aren't just given antibiotics to treat and prevent disease, but to make them grow fatter, faster. But despite a boatload of studies, the agriculture and pharma industries have maintained that the link between antibiotics fed to livestock and antibiotic resistance in humans hasn't been unequivocally proven. In

2018, a study that involved a "molecular clock" analysis would firmly establish that specific antibiotic-resistant genes from *E. coli* cultured from chicken in Arizona farms were precisely the same ones found in supermarket poultry and, subsequently, in hospitalized patients with serious urinary tract infections.

In hospitals, where patients are particularly vulnerable to acquiring antibiotic-resistant bacteria, containment is critical, but conventional infection control protocols fall short of superbugs' capacity to hide, proliferate, and mutate even under watchful eyes. In one study, a team of researchers collected bacterial cultures from fomites, surfaces and objects that can spread pathogens, including bed rails, countertops, faucet handles, and computer mice in occupied patient hospital rooms. They also swabbed the hands and noses of patients and staff, along with the shoes, shirts, and cell phones of staff members. Over the course of a patient's hospital stay, patients' skin and room surfaces became "microbially similar"—meaning the bacteria comingled, cross-contaminating the environment and everything and everyone in it. Based on other reports, the stats are startling: in a three-hundred-bed hospital, there may be up to 64 million possible microbial transfers. Other studies confirm that our surroundings contribute to the makeup of our own microbial communities, and vice versa. Surprisingly, this can include the antiseptics commonly used to scour hospitals and to clean healthcare workers' hands, and chemicals used in antibacterial products such as paints and scrubs, some of which actually *promote* the growth of drug-resistant bacterial strains as they kill off the vulnerable germs and make way for superbugs to flourish. One study even found that bathroom hand dryers can spread antibiotic-resistant bacteria to neighboring surfaces.

Points of transmission also include so-called structural factors, from sophisticated medical devices to IVs and other tubing, which create the equivalent of mass transit for microbes. The year Tom fell ill, a major medical news story would report the discovery that lens maker Olympus, which manufactures medical equipment, knew that its endoscopes could not be adequately disinfected. After several global outbreaks that killed people receiving regular procedures like colonoscopies, the source was traced back to superbugs that hid inside Olympus endoscopes. In Tom's case, some of his doctors speculated that the source of his *A. baumannii* infection was the nasogastric tube inserted in the Luxor clinic to siphon the bile from his stomach to reduce vomiting. We'll never know.

It's no joke that hospitals are now often referred to as the worst place to get well. Superbugs are looking out the hospital windows, licking their chops at the feast that awaits them in this era of superbugs without borders.

Tom: Interlude III

Steff and I are walking in the desert. Not a single living thing is within sight. All I can see for miles is sand, hued a deep red, billowing all around us. The sand stings our eyes and the wind chafes our skin. There is no sun in the sky, but the heat is blazing. We walk in silence, saving our energy. I am so thirsty, but there is no water. My sweat and saliva evaporate into the air with a sizzle. We will die out here, I think.

A Bedouin man suddenly appears as if rising from the dune, directly in front of us. He is dressed in a white flowing robe, and his head is wrapped in a turban. I sense he is a holy man. He walks toward us, carrying a package that he is careful not to drop. He stops in front of us, and deftly unravels ribbons of brown cloth from the package. His skin is tan and supple; his fingernails are buffed oval moons. We eye his package hungrily, like animals. I am hoping it is water, or food. Steff looks at me, thinking the same thing. We know we are not supposed to speak. The man looks up at us. His eyes are black pebbles, recessed deep inside his skull. It is impossible to tell how old he is; his skin is unlined, although his beard is long and white.

The man presents each of us with the contents of the package, which are two small wooden boxes that fit neatly into the palm of our hands. Each box is intricately carved with our names in hieroglyphics. Steff and I open our boxes simultaneously. Inside mine is a tiny green leaf, identical to Steff's. The leaf is plump and waxy and reminds me of a fig leaf.

"Eat," the man says to us. Steff and I look at each other and then at the holy man. Surely he is not serious. How can a single leaf, one so impossibly small, nourish us in a desert, where there is no water? Steff does not hesitate. She picks up her leaf, and gobbles it up. I cannot. My mouth is too parched. I am too tired. I wait.

The man is gone. Vanished. As if he were a mirage. Maybe he is. We walk more, for days, months. Our feet sink deep into the red sand. Each step is more difficult for me, but Steff is walking faster. She is impatient, waiting for me, but she does not say this. The holy man reappears, as if rising from the heat waves, with outstretched hands. He points to Steff. "My child, you have eaten, so you may leave."

Steff does not look back at me; she just drifts away. I see her figure getting smaller and smaller, until she disappears. Now I am alone with the holy man. I am filled with dread.

"You have one day left to eat, or you will remain in the desert for one hundred years," he says solemnly, shaking his head.

I look down at the box in my hand and lift open its lid again. The leaf is even smaller now, shriveled and dry. I pick the leaf up carefully by the stem, but it flakes into tiny fragments. Some blow away in a gust of wind. I only succeed in eating a small piece that remains attached to the stem. I want to cry in despair, but I have no tears left. I want to ask the holy man for water, but he has evaporated. I am overcome with a profound sense of loneliness.

My palm, holding the empty box, is like an alluvial plain. Erosion has weathered my skin, which flakes off like white, scaly scabs, revealing deep red rivulets that carry my blood. The ridges and furrows of my fingers and thumbs are now devoid of whorls, a dermatoglyphic aberration that means I am no longer a person. I am a cadaver. The skin on my feet peels off like snakeskin and falls down, to what I recognize suddenly is the TICU floor. A woman sweeps up my skin with a broom. I am so worthless that my body is being disposed of. I am disappearing cell by cell into a red bin that is labeled with the word BIOHAZARD. *I open my mouth to scream, but no sound comes out.*

I know that this is how I will die.

12

THE ALTERNATE REALITY CLUB

December 24, 2015—January 16, 2016

In no time, Christmas was upon us. The girls flew down from the Bay Area to be with Tom and spent most of the holidays shuttling between our house and the hospital. In normal times, they'd always loved hanging out and watching TV together on visits, when they weren't swimming in the ocean, that is. Now all the action was on the flat screen, but Tom's sagging spirits were buoyed by their presence. And it meant I could spend a few days in Vancouver with Cameron, who was about to turn twenty-three. He'd been born on Christmas Eve, so his birthday had always had the happy holiday aspect to it. As a child, Cam had preferred low-key birthdays at home with his dad and me. He was hell-bent on collecting all available Pokémon trading cards and memorized each form they evolved into. He also played Pokémon on his Gameboy, pitting them against each other and watching them obtain new superpowers after their successive battles. At the time, the Pokémon theme song drove me and other parents insane: "Gotta catch 'em all, gotta catch 'em all!"

For a kid who just didn't "get" other kids (and they didn't get him), he'd been pleasantly surprised at university, where he finally made some great friends. Every year around Christmas, I'd treat Cameron and his buddies to a big holiday dinner that served as a birthday party. He'd met these friends in the campus Alternate Reality Club, and at least this year, it seemed a fitting name for all of us. Only their version was an upbeat take on life; ours was no fun.

Another of Cameron's early obsessions was building Star Wars Lego action figures, which I delighted in, since he and I were both big fans. One day, when he just three, and his dad, Steve, was serving him dinner, I'd reminded him to use his fork. He'd looked skyward, grabbed his fork in his fist and cried, "Use the fork, Luke!" We all burst out laughing—what I wouldn't do for some help channeling the Force from Obi-Wan Kenobi right now.

We had a long tradition now of going to the Star Wars premieres together right before Christmas. Months ago, I'd pre-purchased seats for Cameron, me, and Tom to go see *Star Wars: The Force Awakens*. It was the first Star Wars movie in years, and Cam and I had been counting the days. Obviously, Tom wasn't going to make the movie this year, so Cam brought along his friend Jesse instead. The movie was fantastic, but I couldn't help but compare the fights between the Jedi and the First Order to the one going on in the TICU between the *A. baumannii* bacteria and Tom's immune system. Could Tom's antibodies and natural killer cells annihilate the bacteria? I remembered Chip's PowerPoint that he'd sent me in Frankfurt pinpointing the anatomical ampulla of Vater—the very junction where that gallstone had been unable to pass and where the whole damn pseudocyst had started to form like the Death Star. Could we find some kind of biological blaster to zap Tom's *A. baumannii*?

I was already showing signs of PTSD and didn't know it at the time. Every time one of the X-wing pilots blew up a ship, the thundering Dolby surround sound from the movie theater pounded inside my head. My nerves felt ragged and jangled; every suspenseful scene triggered an exaggerated sense of shock. I looked sideways to Cameron and Jesse to see if they were reacting the same way. Both were sitting back, eating popcorn and grinning, taking it in like a scene in a movie. Because it *was* a movie. But for me, my world was shattering. Part of me was in the movie, and another part of me was watching the other me and Tom getting blown to bits. Afterward, Cameron could see I was shell-shocked.

"Mom," he confided, "maybe you should just go back to watching *Forensic Files* until Tom gets out of the hospital."

I'd been a longtime fan of this corny TV docudrama that we all loved and loathed. Lately, I'd taken to binge-watching it like a zombie, occupying some nail-biter region of my brain while the calm, analytical part continued to work on the puzzle of saving Tom's life, which was like a special edition of a Rubik's Cube. The show was a dramatization of true crimes that were neatly solved with the help of forensic science, all within the program's allotted twenty-four minutes. Somehow, the answer was always right in front of their eyes; they just had to use cutting-edge science and technology to *see* it. In Tom's case, the solutions seemed to come and go: now you see it, now you don't.

Cameron was no blind optimist by nature. His brain, like mine, was hardwired for unsentimental logic and reason. But compassionate qualities had emerged in different ways in recent years, especially with the death of his father several years earlier, and now through this family crisis. Over a beer before I left to fly back to San Diego, he toasted me, telling me that he was very proud that I saved Tom's life. I was deeply moved. Whether it was true or not seemed to change hour by hour.

Tom's clinical status seemed to be oddly in limbo. Although his *A. baumannii* was now resistant to each of the antibiotics that had once looked promising, as long as the bacteria remained contained in the pseudocyst, something appeared to be slowing the bacteria's voracious growth just slightly. It might have been his own immune system, which seemed to rally sporadically.

By New Year's Eve, the mood at home was lifting as Tom continued to make small gains. I could hardly wait to usher in 2016. I wanted 2015 behind us so badly. It seemed like magical thinking to imagine that we could just turn the page. But on the plus side, his condition was now officially listed as "stable"—something that just a few weeks ago we never thought we'd see. And he'd been assigned to the more general hospitalist team to oversee his care, rather than the pulmonary and critical care team. The fact that he was shedding specialists was a good sign. And on New Year's Day, something remarkable happened: Tom was moved from the TICU to a regular ward. The girls and I were ecstatic. Contact precautions were still enforced, but now he would be able to get more sleep and regular physical therapy. The page had turned.

At home, I adopted two new kittens to join stodgy Newt as the welcoming committee for Tom's return home, which I was hoping would be any day now. Bonita and Paradita were both strays from Animal Rescuers without Borders. I took a little video of them chowing down at the same feeding bowl one afternoon and showed it to Tom. He was amused, but it made him even more homesick. As New Year's resolutions go, getting Tom home was the one we all shared. But with every new complication and obstacle, we were forced to reconfigure expectations and then renew that resolve. And wait.

One day in early January, Davey met me for lunch after we visited Tom together. He wanted to fill me in on the results just in from the antibiotic synergy testing that the lab had run. After all this waiting, the findings were a major disappointment. The tests had turned up no new antibiotic approaches—no combination had an effect. Tom was holding on, stable, but we were running out of options to kill the *A. baumannii*. Over a burger, Davey happened to mention a publication he had read by one of our colleague's research labs. It reported that a combination of three antibiotics—rifampin, azithromycin, and colistin—is synergistic against some multi-drug-resistant bacteria, at least in the lab. Normally, no one would use azithromycin to treat a Gram-negative bacterial infection, but the team's research suggested that the azithromycin weakened the bacterial cell wall, which allowed the colistin to get inside, where it could kill it. I was intrigued. That night, I did a PubMed search, found the paper, and read it myself.

A year ago, a study like this about mix-and-match antibiotics would have made for interesting conversation—the kind you discuss at a brown-bag seminar watching colleagues' projects in fascinating areas far removed from your own work. Now, after taking a leave of absence from my usual work because keeping Tom alive consumed me, even a single study that suggested an alternate route out of this dead end was worth tracking down. Could we try this approach on Tom?

There was only one way to find out.

I'd crossed paths with the senior author, Dr. Victor Nizet, at university meetings, and I didn't hesitate to drop him an email now. Would he be willing to test his drug combination against Tom's bacterial isolate? He agreed, and suggested that they first run some tests in the lab to see if the synergy was observed on Tom's superbug. Within a few days, Dr. Monika Kumaraswamy, an infectious disease fellow from Victor's lab, showed that the combination had some benefit, but that was in a Petri dish. Still, it was something.

The next day, I asked Tom if he would be willing to try this new experimental antibiotic combination if the docs okayed it. He was on board. Next, I emailed Chip to see if the rest of the infectious disease team gave the thumbs-up. Chip agreed, pointing out, as Davey had, that there was little harm in adding azithromycin to Tom's regimen because it was a much safer antibiotic than the others he was already on.

On this new antibiotic cocktail, Tom started to improve marginally over the following week. He slept, started to eat soft food, and even kept some of it down. He worked with Amy, the physical therapist, twice a day, learning to sit up and then stand. Victor's antibiotic combo was not a cure, but by holding the infection at bay, it bought us more time.

Chip was encouraged.

"As long as we can keep the Iraqibacter contained in the pseudocyst and maintain adequate drainage," he said, "Tom's immune system can hopefully recover enough that it can kick this infection on his own."

It was the first really good news we had had in weeks. But I was still worried, especially about Tom's mental state. His delusions were getting more frequent and more bizarre. It was hard to know if they were some part of the illness, the mix of medications, the isolation in the hospital, or the way his mind was attempting to cope with it all. Or maybe the answer to this multiple-choice quiz was *E*—all of the above. Whatever sense these nightmares might make internally to his psyche, from the outside they looked like a man unhinged. They were taking a toll on me, too. Every time his delusional dramas spiked, I felt we were on two sides of an abyss, with only a shaky bridge between us that Darth Vader was trying to destroy.

Before dawn one morning, my cell phone trilled Tom's signature ringtone: Thomas Dolby's "She Blinded Me with Science." I had to scramble from bed to reach it on the charger.

"What's going on?" I asked him, shivering, as I crawled back under the comforter.

"Do we have property in a country that doesn't exist?" he asked me anxiously.

What the hell was he talking about? And how do I deal with this?

"Honey," I began. "I don't know what you mean."

Tom sighed. He was clearly very agitated. "Think back, way back. Down, down...to the beginning, when we first met. Did we buy—did we purchase a house in a country that doesn't exist anymore? Have we lost everything?"

This is what he was calling me to discuss at four o'clock in the morning? It was *loco*. I decided to try some logic, keeping my voice calm and quiet so I didn't wake the girls, and with hopes of calming him down.

"Tom, it's the middle of the night. I'll be there in a few hours and before I come, I will check all our financial papers to make sure that everything is accounted for." Deep, cleansing breath.

We hung up, but there was no way I could get back to sleep. I made a giant pot of Peet's coffee in the cone drip filter that Tom insisted on as our only means of brewing it, as he was a coffee aficionado. It overflowed onto the countertop and made a giant mess. I swore loudly. The kittens scrambled under the kitchen table. Newt eyed me and kept his distance.

On my drive in to the hospital, I was deciding how to approach the conversation with Tom. We had to talk about his delusions. Nobody else was. But as soon as I entered his room, I could see that Tom had a sheepish look on his face. He immediately recanted his story.

"Sorry, baby," he said despondently. "I don't know what I was thinking. I guess there are Trojans in my head, just like the song on the radio."

••••••••

To hear Tom apologize was painful; there was nothing he could do about it. And it seemed there was nothing the girls or I could do, either—it wasn't as if we could avoid saying or doing things that might trigger him. He would go in and out of these psychotic states without any provocation. The evidence of his mental deterioration was hard to overlook. He often lay still and practically lifeless in his "recovering" state. Then the delusional dialogues would erupt, like breakthrough bleeding from some parallel universe.

Some days, he didn't even recognize me. Once, when this happened, I decided to take matters into my own hands. Making sure the door to his room was shut, I lifted my shirt over my head and flashed him, yelling: "If I wasn't your wife, would I be doing this?!" Another time, after the psychiatrist asked him if he knew who I was and he shook his head no, I asked him: "You have been married two times in your life. So, here's the question of the day: Am I the old wife or the young one?" Tom took a guess: "The old one." *Womp womp.* Wrong answer.

Somewhere between data and delusions was Tom's reality. The superbug still lurked inside him, but his lab values suggested he was starting to improve slightly, so, every day now, people would assure him that he was getting better. But Tom was sure he was dying. He showed all the signs of a deep depression, too. Not even the girls' continued presence, which usually raised his spirits, was breaking through this dark fog. One morning when I came in to see him, he looked at me gravely.

"We need to call a family meeting," he told me. "To talk about pulling the plug."

"What?!" I cried. "What are you talking about?"

"I'm dying. Face it. Davey and I had a big talk in the middle of the night, and he told me that it's over. So we need to talk about ending it all. Euthanasia."

I was floored. How could this be true? There was no way.

"Tom. Honey. Baby. I just spoke to Davey yesterday and he didn't say anything of the kind. He, Chip and the other docs think you're getting better! Is it possible you could have imagined this conversation of yours?"

I grabbed his phone and looked at the recent call history. It showed that Tom had tried to text Davey in the middle of the night, but there was no sign of a call. I showed Tom the phone. He was adamant that he was right. Exasperated, I called Davey.

"What's up, Buttercup?" Davey replied in his singsong voice. Davey's supply of Southern terms of endearment seemed endless, but always somehow reassuring, as silly as it seemed.

"Plenty," I said. "I am here with Tom right now, and he thinks you and he had a talk about euthanasia last night. He says you told him he's dying, so we need to pull the plug!"

I heard Davey take a sharp intake of breath. "I'll be right there," he replied. "Don't do anything rash."

When Davey arrived, he confirmed that there had been no such conversation the night before. Again, we showed Tom the call history, which showed no call had taken place. Tom was chagrined and shaken by the growing realization that he could no longer trust his own mind about what was real versus what was imagined.

"Davey," Tom told him, "I'm losing my mind."

Davey's throaty laughter filled the room. "Of course," he said, with a lopsided grin.

"No, really," Tom said. "I sometimes see stuff... weird shit..."

Davey nodded. "Tell me about it." Davey had had his share of medical problems, too. A few years ago, he'd spent a few weeks in the TICU after suffering a series of mysterious strokes. The neurologists ran a bunch of tests but never came up with a diagnosis. The experience had given him a deeper understanding of what it was to be a patient.

Tom sighed. "Where do I start? Just the other night, I thought I was crawling on the floor around the nursing station, and the nurses were poking me with needles and scraps of metal. So, later, when I was taken to IR, I thought all of the little metal pieces inside me were going to be drawn out with a magnet—and that it would kill me."

Davey sat down on the corner of his bed and took Tom's hand in his own gloved one. "It's ICU psychosis. It happens to most people who end up in the ICU, or anywhere in the hospital really, for an extended period. You can't tell day from night, so your brain gets mixed up. When I was in the hospital, I could have sworn that I was lying on the floor of my childhood tree house, back home in rural Tennessee. You're not going crazy. I had my doubts before, but it looks like you're actually going to get out of here."

••••••••

One week later, in mid-January, Dr. Gandhi, the hospitalist who was assigned to Tom's case that week, told us that Tom would be discharged to a long-term acute care facility within the next few days. Tom was elated, but I was immediately nervous. Sure, the lab tests showed that Tom's immune system was reviving slightly, that his body had managed to wall off the *A. baumannii*. But no way did he look like a guy on the mend. Most of the time, he "rested" still and pale as a corpse. And it seemed increasingly that although his body had walled off the invader, his mind was walled off, too—from us. There was nothing encouraging about the continuing delusional episodes and periods when he seemed lost in some dark inner world. Where was that being taken into account in the workup for this discharge? Nevertheless, I texted Chip and Davey to give them a heads-up. Both of them expressed concern that it was still too early, but Tom was angry that I wasn't more supportive of the transition. He looked at me with pleading eyes. "Can't I just come home?" But since he was still receiving IV antibiotics, home was not an option.

The next day, an infectious disease doctor who worked with Chip and Davey, Dr. Randy Taplitz, came to see us. I had met her once before, and

Chip had told me that she was one of the best infectious disease docs in the department.

She donned a yellow gown at the doorway and stepped in with a firm yet friendly hello.

"It's great to see you out of the TICU, but we have a hitch," Dr. Taplitz told us. "We cultured a new bacterium from your pseudocyst drain, *Bacillus fragilis. B. frag* is a common gut bug and not typically associated with resistance. There is a possibility it is a contaminant from the drain and is not in the pseudocyst, but it could be a sign of a lurking infection."

"Oh no," I lamented. "Dr. Gandhi told us Tom is getting discharged this week, probably tomorrow."

Dr. Taplitz turned to look at me and narrowed her eyes. "I heard that," she said cautiously. "That's a pretty ambitious timeline. Until we can be sure that this new bug isn't a problem, I'm going to advise that we hold off on discharge. In fact, I would like to put you back on meropenem, Tom, just as a precaution. Mero is an antibiotic you have been on already, and you tolerated it well. Hopefully you won't need to be on it for long." She looked at us for our reaction. Tom sagged back into his bed, but nodded slowly, resigned.

I shrugged. "You know best," I told her. I was secretly relieved. No way did I want Tom discharged to a step-down facility that couldn't handle his complex care. Not yet anyway.

"Good," she said, as she stepped back to the doorway, stripped off her gown and gloves and washed up. "The last thing we need right now is septic shock."

13

TIPPING POINT: FULLY COLONIZED

January 17–February 14, 2016

"Wow, would you look at that!" Dr. Gandhi was surprised and pleased to see Tom grab the rail of his bed and pull himself up to half-propped position, a feat he had managed the day before for the first time since he fell ill, with unrelenting encouragement from me and Amy from the physical therapy team. "You're much stronger now—that's great! Let me just take a look at your belly." Tom scooched to the side of the bed and let his legs dangle over the side while Dr. Gandhi palpated his abdomen and listened to his lungs. He nodded with approval, and Tom inched back to lie back on the bed. Dr. Gandhi had been stopping in each morning to check on Tom, and we'd found some cheerful common ground in conversation about Indian food. A native of Delhi, one of my favorite cities, he and I idly chatted about the experiences I had collaborating on research with heroin users there. Tom chipped in and spoke about the HIV project he had led with sex workers in Nagpur, and his love for South Indian food, *dosas* with *sambal*, *channa masala*, *idli*, and *puri*.

On the other side of his bed, the day nurse, Erin, was entering Tom's vitals into the online medical record system. "So, feeling hungry for lunch?" she quipped. "How about McDonald's—say, Chicken McNuggets?"

We all laughed. Tom was still on a liquid diet and only kept down about half of his meals. And he wouldn't be caught dead eating McDonald's.

"OMG, your ostomy bag is full again," Erin noted with surprise. "I just emptied it an hour ago."

I peered around from my location at the end of Tom's bed. "It's a weird straw color too," I remarked. "It's been brownish-yellow and cloudy up until now."

Dr. Gandhi's brow furrowed. "Keep an eye on that," he said to Erin. And turning to me, he said, "That looks like ascites fluid, which is really strange. I will page the GI resident to get a bigger ostomy bag up here and to take a look."

"It's pouring into the bag now," Erin said quickly with a growing sense of alarm.

"Measure how much you are collecting," said Dr. Gandhi, even more quickly.

"It's five hundred milliliters—and that's just in the last hour," she replied, holding up the ostomy bag to show him that it was almost half full again.

"Should we keep a sample to test?" I asked him.

"Yes. Good idea," he replied. "Let's do that."

"I'm cold," said Tom in a quiet voice. "Can I have a warm blanket?"

"Sure," said Erin. "I'll call for one—"

Suddenly, Tom started to shake. Violently. "I am so cooooollllld..." he whispered. Beads of perspiration hung at his brow and his cheeks were mottled.

"I don't like the looks of this," I told Dr. Gandhi, biting my bottom lip.

"I don't either," he replied. In an instant, concern had turned to alarm. "He's going into shock—hold on." In seconds, he'd paged the TICU resident on call to come *stat*, and Erin used her walkie-talkie to page the charge nurse.

Tom's breathing shifted suddenly to rapid panting. He started to shake so violently, the bed frame rattled. I heard Davey's voice inside my head. *If he shivers so much the bed shakes, that's rigors: a sign of septic shock.* I looked at the cardiac monitor and saw Tom's blood pressure drop from

110/72 to 90/55 in a matter of minutes. His respiration rate increased to 35, then 40 breaths per minute. By now, I knew that his normal *res* rate was less than 20 per minute.

"His BP is in freefall!" I cried out. "And look at his *res* rate. Holy crap!"

Dr. Gandhi was on his cell phone, pacing in the small space between Tom's bed and the window. "Yes, *stat*," I heard him say. "He's gone into shock."

The charge nurse, Julie, came barreling in, pulling on gown and gloves as she and Erin approached the bed. "Should I call a code?" she asked Dr. Gandhi, as she piled several warm blankets on top of Tom. The bed was shaking violently, and I could hear Tom's teeth chatter.

The doctor paused a second to think. "Not yet," he replied. "But get him on oxygen. The TICU resident should be here any second." He looked at his watch.

On cue, in rushed a young doctor that I recognized from the ICU by his spiky short brush cut. Having run from the other side of the second floor, he was out of breath, his white coat flapping behind him. As he paused at the doorway to pull a yellow gown over his head, I read the ID tag that hung from his lapel: DR. WANG, PULMONARY AND CRITICAL CARE RESIDENT. He nodded to me in recognition as he raced over to Tom.

Julie had finished taking his temperature. "One-oh-two point seven," she announced in a curt voice, as she looked to the two doctors for more instructions. I was at a loss what to do, so I ran to the restroom a few yards away to get a cool cloth for Tom's forehead.

"Hang in there, honey," I whispered to Tom, as I wiped his face with the cloth seconds later. "We'll get to the bottom of this. I won't leave you." Tom looked at me and blinked slowly, his eyelids heavy. Behind them, I saw fear.

Dr. Wang finished his assessment within three minutes and was on his cell phone. As I looked over at Erin, I could see that she was emptying his ostomy bag again; it was full. Dr. Wang hung up and turned to the rest of us. "We have a bed open in the TICU. Let's get him over there."

"Now?!" I was in shock myself. I suddenly remembered the dire statistics that people die from septic shock all the time, even in the US. And to be contemplating a code—hospital slang for an emergency of a specific sort—meant that Tom might become one of those data points. Tom knew it, too. As the team readied him for the transfer, he gave me a knowing look that said, *See, I knew I was dying.* I swallowed hard over the growing lump in my throat. What a time for him to be right.

Julie, Erin, and an aide started jamming Tom's toiletries into large plastic bags marked BIOHAZARD. "Help me get his stuff together," Julie ordered me. "We need everything to go with him." Within minutes, Tom was wheeled back to the TICU on his bed, with Dr. Wang and Dr. Gandhi flanking him on either side. Dr. Gandhi pushed the IV pole. I ran along behind, carrying three bags of his belongings, my purse, and my backpack.

En route, we ran into Dr. Taplitz. She stopped dead in her tracks, her eyebrows raised in surprise. I was so relieved to see her. "What's going on?" she asked me. She turned around and together we half-walked, half-ran behind Tom's bed as they wheeled him down the hall, back to the TICU.

I dropped all formalities. "Oh, Randy!" I cried. "One minute he was sitting up in bed, and then... and then his drain started... and still is... pouring out all of this pale yellow liquid, and then he got cold, and now he has a fever..."

Randy's forehead creased. "That sounds like ascites. And he's gone into shock. I wonder if it's the *B. frag* we cultured from his drain. If it is, we're covered. I started him on meropenem last night. But they need to rule out an MI too." I mentally accessed the Wiki page in my brain. Myocardial infarction: heart attack.

The double doors swung open and there we were again. Back in the TICU. Tom was raced to the end of the hall this time: Bed 11. It was a large rectangular room with a window at the end, but Tom was no longer responsive; by the looks of it, he might never see out of it. His eyes were closed and his face was beet red. His breath heaved in short rasps that sounded like a death rattle. A bevy of doctors and nurses started working on him, hooking him up to a new cardiac monitor, repositioning his IV, and taking his vitals. His RR continued to soar to 45, then 50. From the corner of his room, two familiar faces appeared at the door: Marilyn, the charge nurse, and an ex-military nurse, Joe. Both had been with Tom through so much—and now this.

"What happened?!" asked Marilyn. "We just saw you the other day and you told us Tom was getting discharged soon."

"That's what we thought, too," I replied dejectedly, hugging myself. "But not now."

Joe put his hand on my shoulder. "Hang in there, kid," he offered lamely. Joe had wise, blue eyes and a short brush cut. He didn't take any bullshit from Tom on those days that he had been up to dishing it out. But he and Marilyn, both clearly shaken now, retreated back to the nurses' station to give the critical care team more space.

Randy stayed with me in the corner of the room as the TICU team completed their assessment, thank God. I was so grateful for the support. Chip and Davey had told her that I didn't have any other family living in San Diego. I felt like I was having an out-of-body experience, looking down at Tom, the doctors and a woman with dirty blond hair who looked like a shell of a human being: me.

One of the critical care doctors, Dr. Mims, barked an order I did not immediately understand. "Page anesthesia, *stat*!"

I turned to Randy. "Why are they paging anesthesia?" I whispered to her.

"They need to intubate him," she explained calmly.

"Oh no, not the ventilator! Oh God…" My voice started to escalate. It didn't sound like me. It sounded like someone who was becoming hysterical. My inner voice whispered: *life support*. Translation: *death*.

"It's gonna be okay," she reassured me. "Trust me. He needs the support to breathe right now. Hopefully it will just be for a few days."

Within minutes, a new team of doctors arrived. There were now more than ten in Tom's room. They gathered around his bed in a swarm, so many that I couldn't even see him anymore. Tears slipped down my cheek one by one, then in steady rivulets. Someone handed me a Kleenex. I heard someone else say that three liters of fluid had now poured into Tom's ostomy bag.

A trim blond woman that I didn't recognize emerged from the swarm and approached me. "Mrs. Patterson?" she asked me.

I looked at her blankly. Honestly, in the moment the name didn't register. "Uh, yeah. Steffanie Strathdee, actually. But yes, I'm Tom's wife."

She handed me a piece of paper in her gloved hand. "I'm Dr. Meier, the anesthesiologist. We need to intubate your husband immediately to help him breathe. He probably won't remember anything about this and we hope he doesn't, since the memories can be . . . unpleasant. Can you provide us with consent please?"

I nodded and signed the form. She thanked me and returned to the swarm, directing the activities of the hive. Within another five minutes, they all withdrew from the bed, their job done. I gave a sharp intake of breath. Before me, my husband had a face-hugger strapped on, through which a giant tube entered his mouth, like the man who had died in Bed 9. It was connected to a contraption the size of a small child that took up all the space to the right of his bed. It had several gauges and dials that I later learned were set to max. The dreaded ventilator. *The vent.* With a start, I realized that Tom could no longer talk. And that I couldn't even remember the last thing we talked about. Would I ever hear his voice again? How did this happen, and so quickly? I felt totally disoriented.

Randy looked at me and her eyes narrowed. "Are you okay?" she asked me.

"Yes. No. I don't know, actually." I looked at her and tried to smile. And failed miserably.

She handed me another Kleenex. "Do you have anyone around that you can call on, that can help give you some support?" she asked me gently.

Both the girls had been down in San Diego for half of December and

January, and Carly had just left a few days ago to deal with, among other things, the burglary of her house during the holidays. We were all making decisions day by day and doing the best we could.

"Tom's girls *just* went back to the Bay Area," I told her vacantly, hardly able to focus. My good friend Liz was out of the question. Although she only lived a few doors up the street, her husband had just been diagnosed with end-stage pancreatic cancer. This was no time to add to her burden. So many other people were supporting us in so many ways, but...

"Call the girls," Randy said. "I can talk to them if they need more information," she offered kindly. I confessed that I hesitated to call them because I knew I was getting a reputation for crying wolf—overreacting to the point that they had a hard time determining how serious the situation was at times. I didn't blame them. I had a hard time gauging the gravity of these dips and dives myself. I was not naturally an alarmist. But how was I supposed to tell that a crisis would resolve, as, somehow, remarkably, they kept doing with Tom?

Randy assured me that I couldn't possibly overreact under the circumstances. Not this time.

••••••••

The next twenty-four hours were touch and go. Tom was placed in a medically induced coma while the docs worked to get his sepsis under control. He was given extra fluid to replace what he lost and sent for a CT to determine what was going on in his abdomen. His hemoglobin dropped precipitously, and he got a transfusion. *Three pints of blood.* He had lost at least one quarter of the blood in his body. While I waited for the results of the CT, Davey arrived. I ran to him and he gave me a big bear hug, and I buried my face in his neck.

"Hi, Buttercup," he said quietly. "I'm up to speed. I stopped by radiology to look at the CT."

"And?" I asked him anxiously.

Davey looked grave. "Tom's pseudocyst drain slipped and dumped all of the crap inside it into his abdominal cavity. This is what I was afraid of," he said. "The *Acinetobacter* is now everywhere in his body. He is now fully colonized."

"Oh my God," I whispered. It was the worst possible news. "And we have nothing except Victor's cocktail of antibiotics to treat it, right?"

Davey shook his head and tried to meet my gaze, but couldn't. "That cocktail is bacteriostatic, keeping the infection from advancing, but it isn't bactericidal—"

"Meaning it can't kill it," I interjected.

"Right," Davey replied softly. "It's no match for the *Acinetobacter*, especially now that it's spread everywhere. We took samples of his blood and his sputum, but I bet my white coat it's going to confirm *Acinetobacter*."

My eyes filled with tears. "Is there any good news at all?"

Davey thought for a moment. "Well, there's no sign of an MI. And Randy probably saved his life by putting him on meropenem yesterday. The *B. frag* was definitely growing in his blood first. Without the mero, he probably wouldn't have made it this far."

••••••••

Four days. That's how long Tom "slept," kept under with propofol, the "Michael Jackson drug" that its creator called "milk of amnesia." Each day, he was wakened briefly to make sure that he was still able to hear and follow commands by giving a thumbs-up or nodding his head. The day finally came when it was time to turn the propofol off to try to wake him up for good. He emerged from that netherworld slowly over the next few hours. Wiping four days of crud from his eyes, I could feel the warm blood rush to his cheeks as I touched his face with my blue-gloved hands. He finally looked... alive.

"Hey, Rumpelstiltskin," I whispered. "Welcome back to the land of the living!"

Earlier, when I'd posted an update to Facebook, sharing with friends that Tom was in a medically induced coma, I'd invited any suggestions to add to our improvisational music therapy. They poured in with an eclectic mix: Leonard Cohen, David Bowie, Lucinda Williams, Timber

Timbre. The girls and I made playlists on Pandora that Tom could listen to around the clock, assuming he could hear. I Skyped daily with Robert, my spiritual rock, and the girls and I stayed in touch with Martin, a holistic healer in San Diego who shared a special place in the family and often came to sit bedside and lay his hands on Tom. We received messages from around the globe that friends and colleagues were lighting candles and saying prayers; the outpouring was overwhelming. I had no idea if any of these things would work, but I figured they couldn't hurt. In a struggle that was so isolating in so many ways—we still couldn't touch Tom without protective gear—it seemed that he and we were at a vortex of caring, loving, healing energy and presence. It was like nothing I'd ever experienced before. Tom, still on the vent and unable to speak, seemed oblivious to most of it, but at times, someone's presence roused him from the deep or coincided with a positive shift in some of the monitors tracking his heart and other vital signs. But only sometimes.

I never knew what a day in the TICU was going to hold. My routine call at five a.m. to get the report from the night nurse was my feeble attempt to get a handle on a situation that I could never have control over. Each day that I made the half hour drive to the hospital, I asked myself, *Will Tom be conscious or unconscious? If he is awake, will he even recognize me?* The days that he didn't were the hardest; it made me feel like my presence was totally useless. Deep inside, I knew that wasn't true, but it was getting harder to keep my chin up.

Finally, after several days, Tom spent more time conscious than unconscious and was weaned off the vent. A speech pathologist worked with him so he could start to relearn how to speak. One morning I approached his bed just as he was waking up.

"Good morning," I said to him with a smile. "Do you know who I am today?"

Tom looked at me thoughtfully and croaked, "Me... me..."

I got ready to be disappointed. Note to self: don't ask a question when you are not prepared for the answer.

"Me, what?" I asked him anyway.

"Mi amor..." Tom croaked, and blew me a little kiss. My heart melted. I wished every morning began like that. More often, his fragile grasp on life and his thin defenses against the deadly *Acinetobacter* within dictated darker moods and more startling turns.

One morning, after Tom's physical therapy session, he had a surprise visitor. Bob Kaplan was his childhood friend, a fellow surfer, and lifelong conspirator in manly mischief. Bob had shifted careers from academia to government work some time back and lived in DC, a big shot now, but was in town, had heard about Tom, and stopped by to see him.

"Some people will do anything for a little sympathy, huh, Leroy?" Bob teased him, calling him, as he always did, by his middle name. Tom tried to respond, but couldn't catch his breath.

I lamely reminded him of the tip the nurses used to help him remember to breathe in through the nose and out through the mouth: "Smell the rose, blow out the candle."

"Fuck that," Tom snapped irritably. "Easy...easy for you to say." I stopped hovering and the two resumed their mostly one-way conversation. Bob reeled off one-liners, a little worried when Tom seemed unable to parry in his usual way. Any pretense of conversation ended abruptly when Tom's cardiac monitor alarm went off. His *res* rate had spiked over 30. Will, the respiratory technician, tried every option to stabilize it and clear the goo from Tom's airway, but finally shook his head as the alarm sounded again.

"Sorry, man," Will said, shaking his head. "Gotta call Ghostbusters." And just like that, Tom was back on the vent. This time, they did a tracheostomy—cut a hole in his neck to attach the breathing tube directly to his windpipe—and when they were done, the metamorphosis was complete. He now looked more like a space alien than anything human.

****

Every skirmish threatened to be Tom's last. He'd already survived three bouts of septic shock, and one was enough to kill many people. With every success, the good news was that he survived to fight another day.

But despite each battle won, it felt like we were losing the war. And for good reason: we were.

In recent years, more and more medical journals and media headlines had been reporting on cases in which people were getting sick or injured and developing infections that once would have been treatable with antibiotics but were now fully antibiotic-resistant. While doctors were struggling to treat the casualties case by case, the battle had spread far beyond them.

Left unchecked, an estimated 10 million people were going to die from superbug infections each year by 2050. The former director-general of the World Health Organization, Margaret Chan, had recently stated that we were on the cusp of a post-antibiotic era, where a simple scrape could lead to limb amputation or death. It sounded extreme until you looked at Tom, who was, in fact, dying little by little each day. My husband could soon become one of the more than 150,000 people who die of a superbug infection in the US each year.

The pharmaceutical industry's rote assurances that new antibiotics were on the way ignored the reality that millions of people could die in the years needed to develop new drugs and move them from lab experiments to clinical trials. This speedy antibiotic pipeline was a pipe dream. The truth is more complicated.

Colistin, the last-resort antibiotic that Tom's *A. baumannii* was now resistant to, has been around since World War II, and the last new class of antibiotics was discovered in 1984. Within those classes of antibiotics, a certain amount of tweaking could create new generations of a drug, but bacteria eventually developed resistance to them. Colistin, specifically, was increasingly proving ineffective against bacteria it used to snuff out, not only in the US but globally.

To make matters worse, powerful antibiotics like colistin wipe out the friendly bacteria that help keep our internal microbiome in balance. The temporary loss of these good bacteria is considered part of the routine collateral damage of antibiotic treatment. It's not a critical loss for most healthy people under ordinary circumstances because the normal balance of protective flora typically returns. But if friendly bacteria are annihilated in someone whose immune system is already severely weakened, like Tom's, a superbug like *A. baumannii* can more easily move in for the kill. And each new wave of the aggressive bacteria is more likely to have genetically adapted to tailor resistance for survival.

In increasing numbers of cases, a serious underlying illness brings patients to the hospital for treatment, but what kills them is an unrelated infection they acquire there. These so-called nosocomial or healthcare-associated infections are a growing problem, with governmental estimates that on any given day, about one in twenty-five hospital patients has at least one such infection.

Sick people are also especially vulnerable to the side effects of the antibiotics on other organs and systems. Colistin is considered a last-resort or "salvage" therapy because it can be toxic to the kidneys and nervous system. In its favor, colistin has a high cure rate if the treatment duration is short and the patient isn't already suffering from shock or severe malnutrition. But Tom was suffering from both. And he had already been taking colistin for a month. He was also still receiving meropenem and tigecycline, two other big-gun antibiotics. Tom's *A. baumannii* was resistant to them, too, but taking him off all antibiotics would feel like we were giving up. If there was even a slim possibility that one of them might kick in, we wanted to give it a chance.

Antibiotics or no antibiotics, it was also increasingly clear that this was now much more than a battle against this superbug alone. It was just as much a struggle to survive the cascade of complications—a euphemism, really, for the many ways that the embattled body becomes overwhelmed and begins to fail. One complication and a medical intervention to fix it often lead to another. In Tom's case, as in many, CT scans were essential for an accurate view of a problem area, yet with every scan came the risk that the imaging dye would further damage the kidneys and lead to kidney failure. And although the nurses flushed the catheter tubing from his abdominal abscess regularly, the tubing kept getting clogged and the infection kept spreading. Their solution was for interventional radiology to keep adding more drains; surgery was still considered out of the question. Every procedure added new risks of complications and more sepsis.

Lungs, heart, kidneys, brain. As they falter, it's like watching an expanse of city lights against the night sky, block by block flickering off until all that's left is the dark. Some days Tom could follow the nurses' commands to move a foot or squeeze his hand. Other days he could barely open his eyes. Tom was quickly becoming the poster child for the dystopian future of the post-antibiotic age.

••••••••

On Valentine's Day, I decided to shake things up a bit. I put some lace undies on under my sundress and made a sign that said VALENTINE'S INTERVENTION IN PROGRESS: ENTER AT YOUR OWN RISK. I drew a large heart around the words. It wouldn't compete with the other sign on the sliding door to Bed 11, which was bright green and read STOP! CONTACT PRECAUTIONS. Looking in the mirror above the sink, I applied bright red lipstick and puckered, giving the paper a big kiss right in the middle. "Ha," I whispered to no one in particular, "that ought to do it."

I'd borrowed tape from the nurses' station to put up the sign and had told the day's charge nurse, Marilyn, that I was about to give Tom his Valentine's Day present. When I told her what I planned to do, I'd promised it would be all-look, no-touch to adhere to infection control procedures.

"This is a first for the TICU," she chortled, as she tucked a lock of her Doris Day hairdo behind one ear.

In Tom's room, I shook the mouse on the computer and clicked on YouTube. In the search box, I entered the words "Marcy Playground" and found what I was looking for: "Sex and Candy" was one of Tom's favorites. I laughed out loud. He loved to sing this—or *had*.

"This one's for you, babe," I crooned. Curtains pulled, standing bedside, I pushed the Play arrow with my blue-gloved finger, then carefully pulled my protective gown, and dress, aside for a black lace reveal. No response. Dress and gown back in place, I flipped to our Pandora playlist and took a brush and combed his hair, thinking that I saw him grimace as I pulled at a few knots of matted hair near the tubes that hooked over his ears. Later that morning, little had changed but our place on the playlist. The haunting familiar lyrics to "Hotel California" wafted through the room.

I closed my eyes. How old had I been when I had learned to play this song on my acoustic guitar as a teenager? I hummed along with Don Henley, who sang about trying to kill the Beast with a steely knife. Tom's eyebrows furrowed slightly, or was it my imagination? Whatever he might be hearing, including his favorite music, it was hard to know what his brain might be making of it. Even ordinarily, a quirk in his sensory wiring mixed sound with colors—a condition called synesthesia—so that everything, from Beethoven to the beeps and hums of the bedside monitors, brought a shock of color with it.

Suddenly, Tom opened his eyes and looked at me solemnly. Back on the vent now, with a wide breathing tube directly attached to his trachea, he could no longer speak. But he didn't have to. I knew what he was thinking as I gripped his hand, and together we listened to the Eagles' last cryptic verse. The sense that Tom could "check out" any time but might never leave this place hit entirely too close to home.

Tom: Interlude IV

I am in a terrifying world that no one else can see.

And I am untouchable.

A sign says: INFECTIOUS AGENT.

I am the pariah.

PART III

The Perfect Predator

When you have exhausted all possibilities, remember this: you haven't.

—*Attributed to Thomas Edison*

14

THE SPIDER TO CATCH THE FLY

February 16–20, 2016

Throughout Tom's illness, Davey had encouraged me to call day or night, explained medicalese to me in layman's terms, and didn't sugarcoat the truth. Ever. And I needed a reality check. If I'd been protecting myself from some truth about Tom's situation, then the recent sobering conference call with my colleagues on retreat had ripped off the Band-Aid. I texted Davey and we met for lunch the next day.

I told him about the conference call and how, after I'd described Tom's situation in some detail, they'd gone unusually quiet, then sympathetic before saying our goodbyes. And how in that moment when they thought I'd hung up, but I hadn't quite, I'd overheard the former university chancellor's comment to those around the table, the question I wasn't supposed to hear.

"Has anyone told Steff that her husband is going to die?"

You might think that two and a half months with Tom at death's door would have left me with no doubt that he was dying. But my experience was also that he'd been "dying" for two and a half months—and hadn't died. He'd just kept pulling through over and over again, hanging in there like the Energizer Bunny, just *going and going and going*, waiting for somebody to come up with the right antibiotic.

"Give it to me straight," I told Davey, telling myself to hold it together as I shoveled a forkful of quinoa salad into my mouth. "Tom is dying. He is slowly slipping away. Am I right?"

Davey sipped some Diet Coke and rubbed his day-old peach fuzz. He was thinking how to respond to me truthfully as he toyed with his straw.

"I think so, yes," he said slowly. "Although he is not in multisystem organ failure. He is on the vent, so his lungs are failing, and he is on pressors to keep his blood pressure up, but while his kidneys have taken a hit, he doesn't need dialysis." Davey's voice sounded professional, but his eyes fluttered a little too rapidly. I knew he was blinking back tears, being brave. For my sake.

"Doesn't need dialysis—*yet*?" I asked him, inwardly pleading that he would say, "No, never. Tom's kidneys will not fail." He did not say this. His silence and sad eyes said otherwise.

It was a helluva way to celebrate Tom's birthday on February 18, but we gave it our best shot. My mom and dad flew in from Toronto, and they busied themselves tying balloons to the rails of his bed. Tom was on the trach vent, so he couldn't talk, but he opened his eyes from time to time. It seemed possible that he was at least semiconscious in that haze, and when I saw his gaze land on a balloon, I took that as a cue.

"Happy birthday, honey! You are sixty-nine today!" I held my laptop in front of him and clicked on a video of our departmental staff and students singing "Happy Birthday" and clapping. We sang along. Tom's gaze shifted and he stared at the ceiling. If he could have had a cake and candles, I know what his birthday wish would have been: *Get me home.* We all had the same wish. I couldn't imagine what he was thinking—might be thinking, or hallucinating, for that matter—but if in some rational corner of his brain he was doing the math, he'd realize that he'd spent almost three months in the hospital and there was no sign of him getting better.

On the chair, there was a gift bag waiting, with our names handwritten on a small accompanying envelope. Inside was a card with a hand-painted rainbow and handwritten note with a delicate flourish: "To the most inspirational couple we know, who has taught both of us what true love is really about. Love ML and Yaz." Tom and I had had to miss the wedding of our postdoctoral fellow, Maria Luisa, whom we affectionately called ML, but despite all they had going on, they'd sent this along. In the bag were two white T-shirts, one large and one 2XL. Both shirts had a magnified photo depicting chains of short bacterial rods, with the words I SURVIVED IRAQIBACTER! emblazoned on it. I laughed with delight, draped Tom's T-shirt across his chest, and snapped a photo with my cell phone. He was far from being 2XL anymore, but at least the survival message was encouraging.

"Honey," I whispered in his ear. "You aren't going to win any fashion shows with this, but let's hope that this is the 'before' photo, and in the 'after' one, you will be wearing this instead of your hospital gown the day we get you the hell out of here."

Tom didn't stir. His eyes had closed.

Party over, balloons gone from Tom's bed rails, the room sank back to pallid gray tones. The color of hopelessness. It wasn't like Tom to give up, but how much longer could he hold on? How much fight could he possibly have left? I pulled up a chair and sat and stroked his cheek. He closed his eyes. The Beatles song "While My Guitar Gently Weeps" drifted from the Pandora station. It had always been one of our favorites.

"Have I told you today how much I love you?" I said softly to him, and I saw his head move, just barely.

I wasn't sure, as I climbed into my Prius to head home, how much fight I had left in me, either. I propped my phone in the dashboard cradle for a drive-time chat with my sister Jill in Toronto. Three years younger than me, Jill had always been a glass-half-full kind of person. Now an elementary school teacher, she brought her trademark equanimity each day to a crowded class of fifth graders—and lately to me. She did yoga and meditated, and had developed a new sense of calm—mindfulness before it was trendy—that I admired, especially since I'd never been one to sit still long enough to get mindful. She had always looked up to me, though, for being kind of fearless, or at least that's how it looked to her. It had been hard to show her how vulnerable I felt some days. Today was one of them, and I was too spent to pretend otherwise.

"We're up against the wall, Jill—no treatment options left—and I'm just so, so tired." I had to wipe my eyes to see the road. "I'm not even the sick one, and I'm about to crack."

She listened patiently as I relayed the details of the day and the crushing helplessness that was overwhelming me. In my mind's eye, I pictured her sitting in lotus position on her couch, twirling a lock of her blond hair, like she always did.

"It must be so hard, Steff, and I can see why you'd feel that way," she said. "But think of the worst things that have ever happened to you. You made it through then, and you can make it through this."

She knew better than anyone. She knew my history of being a bullied kid—the nerdy girl who didn't know how to play dumb. I had even been lit on fire one day as a prank. I was about ten and walking home from school one winter day in my favorite shearling parka, when a bunch of boys ran up and dropped something into my hood, and ran off, laughing. The girl I was walking with screamed, "You're on fire!" I didn't even realize it until she screamed again. Instead of panicking, I dropped to the ground and rolled in the snow, then shook off my coat before it could burn my skin.

"And do you remember what you said when everyone asked you how you knew to drop and roll? You said, 'I don't know; I just did what I had to do.' That's who you are, Steff. That's what you do."

By the end of our conversation, I felt the old sense of fearlessness rising. That Little-Girl Me had her long blond hair burned off but still walked the rest of the way home for lunch. And as hard as life was now, I still remembered the night Tom proposed like it was yesterday, and the bioluminescent tide. The glow looks so light and otherworldly, but it's the stress of crashing in the surf that activates the microorganism's bioluminescence.

I knew that feeling. The stress of crashing, anyway. Now I just needed to get my glow on.

"I guess it's time to put on my big-girl pants," I told Jill. "Game on. This bug has messed with the wrong epidemiologist."

Home to the kitties and the mail. Then an early shower, exhausted. A glass of wine. Okay, two. But I couldn't stop hearing that conference call question: "Has anyone told Steff that her husband is going to die?" Well, they have now. But how to tell if someone is *really* dying? I thought back to the late 1980s, when I had volunteered at Casey House hospice in Toronto. I remember reading some of the brochures that the staff shared with families to prepare them for the death of their loved one. Extreme weight loss and muscle wasting. Check. Sleeping most of the time. Check. Loss of cognition. I thought of Tom's continuing delusions and mental deterioration. Big time.

But I'd grown accustomed to the internal debate, the one in my own mind, between the hyperrational Scientist Me, the problem-solving Pit Bull, and the Wife Me, anxious and hoping for a miracle—someone or something that would show up just in the nick of time and save Tom. Maybe I was in denial. Maybe I did need to wake up and accept what was obvious to everyone else. To anyone with a rational grasp of the facts. I thought of my PhD adviser, Dr. Randy Coates, a medical doctor turned epidemiologist who had died at age forty-two. The night before I defended my doctoral dissertation, I had a dream where Randy was quizzing me relentlessly on the phone, even though he had died two years earlier. Each question that he asked in my dream was one I received the next day from my thesis committee; I remember smiling ear to ear and answering each one with a new confidence. I passed with glowing reviews. That was just a dream, of course. But what would Randy say if he was here with me now? *You're stuck on the wrong question. The operative question isn't whether you know that he's dying. It's "How do you save this man's life?" Who cares if you're not a doctor! You're a scientist, for god's sake—think like one!*

Okay, then. Ordinarily, if I were designing a research study, I'd identify the problem to be solved, research the existing literature, track down the top experts in the field, pull together a team, and tackle this together.

I'm an epidemiologist involved in global health research—that's what I do every single day. So, what's stopping me now? I didn't have the answers—nobody did. But I sure knew how to look. And even if looking was all I could do, it was somewhere to start.

Dressed for work—bathrobe and cat leg-warmers for the night shift—I spent the next few hours on the internet, desperately looking for answers. I had no idea what I was looking for, but I had to start somewhere. I entered the search terms "multidrug resistance," "*Acinetobacter baumannii*," and "alternative therapies" into PubMed. There was the 2010 article I'd seen earlier referring to the "significant challenges" that antibiotic-resistant *A. baumannii* presents. The conclusion hadn't been promising, either. Despite the prevalence of *Acinetobacter* infections and the interest in finding effective treatment, there was a dearth of data providing evidence-based options for treatment.

I continued to search, though, and within the hour, I found an article from 2013, published in the journal *Trends in Microbiology*. It was titled "Emerging Therapies for Multidrug Resistant *Acinetobacter baumannii*." The abstract mentioned several alternatives to traditional antibiotics that had activity against *A. baumannii*: phage therapy, iron chelation therapy, antimicrobial peptides, prophylactic vaccination, photodynamic therapy, and nitric oxide–based therapies.

In all the conversations and medical literature about antibiotic resistance that had come up since Tom had fallen ill, I'd never heard the docs talk about any of these. I downloaded a PDF of the paper for a closer read and saved it in a folder on my laptop called "Unconventional Cures for TLP"—Tom's initials.

One by one, I looked up these approaches. A quick lit search determined that iron chelation therapy and antimicrobial peptides had only been studied *in vitro*, which means in the lab, and neither had yet been clinically tested *in vivo*, in a living organism. Vaccination was years away, and photodynamic therapy and nitric oxide–based therapies could only be used topically, on the skin. That left only one treatment, phage therapy—now *that* was interesting. This was the treatment of bacterial infections with bacteriophages, viruses that attack bacteria instead of people.

I sat back on the couch, absently petting Bonita, one of the kittens, who lay on my lap. I closed my eyes, recalling what I had learned in my university microbiology courses back in the mid-eighties. Bacteria are considered the smallest living things on earth, single-celled microorganisms that average about 1,000 nanometers long, or about 0.01—one one-hundredth—the thickness of a single sheet of paper. They are incredibly adaptable and have been found everywhere from inside rocks beneath the ocean floor to volcanic vents. Depending on the species, they can thrive on their own or inside other organisms (as they do in us), sometimes attacking and sometimes coexisting peacefully with their host. They "eat" by metabolizing nutrients in their environment and reproduce by dividing themselves in two. Not sexy, but very efficient.

Phages, on the other hand, are viruses and as a group are notoriously misunderstood, underappreciated, and often maligned, even in the world of science. It's easy for that to happen when the headlines everyone sees are all about the killer viruses—HIV, Ebola, smallpox, influenza, for instance—or pesky common cold viruses. But there are an estimated 380 trillion viruses in our body, making up our virome. These include billions of peacekeepers, phages that quietly go about their work munching on bacteria, maintaining the balance of power among the ranks of organisms in our various microbiomes. Also, because viruses are *so* small, one-hundredth the size of a bacteria, scientists couldn't even see them with light microscopes. And what humans can't see, they have a hard time conceiving of, much less understanding. What earlier lab scientists could see, however, was that their Petri dishes containing thriving bacterial colonies were sometimes ruined by the sudden appearance of clear spots and streaks. Something was killing them off.

I'd seen that myself in virology class, only by the 1980s we knew what we were looking at, even though we couldn't see phages with a light microscope. In a lab assignment, we streaked a bacterial culture onto Petri dishes overlaid with agar, that seaweed-based gel mixed with chicken broth. After the bacteria fed on this solidified agar and multiplied into visible polka-dotted colonies a few days later, we pipetted drops of sewage samples onto the dish, labeled them carefully, and incubated them at body temperature. Within another day or two, some of the Petri dishes looked like Swiss cheese; the holes in the agar were *plaques*—evidence that phages had been hard at work destroying bacteria. Our professor, Dr. Mounir AbouHaidar, inserted the tip of his pipette into a plaque where the phages had attacked the bacteria. Here, he explained, the phage plaques could now be plucked out and added to a flask with billions more of the same target bacteria in warm broth, where the phages would multiply into billions within a few hours. One phage researcher called them "nature's ninjas."

The word *phage* comes from the Greek term *phageîn*, meaning to eat or devour. Bacteriophages are a specific kind of virus that "eat" bacteria by injecting their DNA into them and turning them into phage manufacturing plants. In the process, they destroy bacteria from the inside and cause it to "lyse" (break open), releasing up to hundreds of new phages called virions. Thus, technically, phages don't really ever eat in a conventional sense, nor do they reproduce sexually or in any of the creative ways that bacteria, yeast, or other living organisms do. But they do multiply, and with an astonishing efficiency that puts bacteria to shame—or more accurately, to death.

Phages come in many varieties, but thus far, perhaps the most-viewed and best-documented, or scientifically "characterized," phages are the T4-related coliphages (they attack *E. coli*), which look like tiny alien spaceships with long spindly legs, not unlike the Star Wars Imperial walkers Cameron used to make from Legos. As with many viruses, the "head" of most phages is a protein shell called a capsid, which sits atop the tail and spindly legs, structures that are unique to phages and come in a variety of shapes and sizes. The capsid is usually the shape of an icosahedron, a twenty-sided structure that resembles a geodesic dome—like Walt Disney World's EPCOT Center.

Most phages have hollow, short or long tails that they use to latch on to the host bacteria's cell wall and, like a syringe, inject their genetic material inside. Technically, all of these terms—head, legs, tail—are the language we use because it's easy and familiar, a human-friendly way to describe them. But the vital active core of their being isn't in those exterior features we associate with living things at all. Those external features are just a means of transportation—a disposable nano-size space shuttle for the ribbons of genetic material that ride inside the capsid.

Depending on the genetic makeup of a phage, however, some so-called temperate phages invade a bacterial cell and then have the option to ride quietly along, basically snoozing as they integrate their genetic material into their bacterial host's, until the time comes when an encoded trigger rouses them and they shift into ninja mode. This could be hours or eons. If you're hoping to wipe out a bacterial infection quickly, you want the lytic kind that goes from zero to phage-rage fast.

I had no idea that phages could be used to treat bacterial infections in people, but the idea was brilliant. I read more, hungrily. Where could I find a doctor who could carry out phage therapy? My hopes were dashed within minutes, because I turned up zip. Although the Environmental Protection Agency had approved a phage preparation as a pesticide to treat tomato rot in 2002, and in 2006 the Food and Drug Administration had approved a phage cocktail for the food industry to disinfect meat and poultry of *Listeria* bacteria before selling it to consumers, they had yet to approve phages as a treatment for human bacterial infections. I found an article about a research trial in Europe that was using phages to treat burn victims, but the NIH's clinical trials website didn't indicate that there were any ongoing trials anywhere in the US. My online search found no phage treatment protocols that we could apply to Tom's *Acinetobacter.*

This was puzzling on two fronts. First, I found articles on phage therapy dating back to the 1930s and '40s. There were case reports of its use to treat *Salmonella* infections, published in one of the world's most prestigious medical journals, *Journal of the American Medical Association*. Second, my Google search quickly identified that phage therapy was being used regularly in some former Soviet bloc countries, like the Republic of Georgia, Russia, and Poland. A *Buzzfeed* article from March 2014 described the desperate attempts of patients with superbug infections who had flown to Eastern Europe, where the Eliava Phage Therapy Center in Tbilisi, Georgia, had existed for decades.

In most countries, phage therapy had fallen out of favor after penicillin came to market in the 1940s. Understandable, since antibiotics were true miracle drugs until antibiotic-resistant bacteria began to surface as a significant concern in 1959. The alarm over a coming pandemic of deadly antibiotic-resistant bacteria threatening human life had been sounded the world over with no apparent effect. Outdated ideas, ignorance, and plain prejudice had hobbled the scientific and medical community. Although phages were studied extensively by basic scientists in molecular biology and genetic engineering experiments, relatively few scientists had persisted to research phage for its therapeutic potential, most of them laboring in obscurity at universities or a few small biotechs.

As much as some phage therapy proponents were eager to see it embraced as a potentially "new" way of treating antibiotic-resistant bacteria, bureaucratic hurdles and the lack of empirical data on efficacy had thwarted attempts to advance its clinical use in the Western world for more than a century, and that muddled past had muddied prospects ever since. Phage therapy clinics in Tbilisi and Wrocław advertised treatment and there were online accounts of some success stories, but few rigorous studies in humans had been published in English journals.

I could not find a single article describing phage therapy to treat humans infected with *Acinetobacter baumannii*. It had been done in Petri dishes. In mice. A few rats. It seemed promising, but could I really justify turning my husband into a guinea pig? If things went sideways, how would I explain my decision to inject him with a legion of viruses to his daughters?

Using viruses to chase bacteria reminded me of the children's song about the old lady who swallows a spider to catch a fly:

She swallowed the spider to catch the fly
But I don't know why she swallowed the fly—
Perhaps she'll die.

••••••••

It was after eleven p.m. and I was starting to fade, but heard a ping from Facebook messenger—my colleague Maria Ekstrand in San Francisco who knew both Tom and me well wrote to tell me about a friend of hers who had flown to Tbilisi to have her MRSA infection treated at the Eliava Phage Therapy Center. It had worked. Another cosmic coincidence? Some might call it a sign, and maybe it was, but I needed more than a sign. I needed a path forward, even if I had to build it myself. I fired off an email to Chip, attaching a research paper on *A. baumannii* phages I had found by Dr. Maia Merabishvili, a phage researcher from the Eliava Center who was now based in Brussels. I could just imagine Chip's eyebrows twitching as he read my email.

Dear Chip, I know that we are running out of options to save Tom, so I have been exploring alternatives to antibiotics. What do you think about phage therapy? I know it sounds a little woo woo, but it might be worth a shot.

If I was going to try to obtain an experimental treatment to save Tom's life, it was going to take an act of God, a lot of luck, and more energy than I might be able to muster. But just thinking about phage therapy gave me an adrenaline surge. I'd loved my undergrad virology class years ago, and now the challenge was anything but academic. I couldn't help but find it profound that I had to return to my own past, my training, to bring an obscure thread of science forward to find a cure for Tom. And that the potential answer could have been there all along without me or anyone seeing it until now—I could feel the excitement building in my gut. And it wasn't the knot of fear that I'd been learning to live with since Luxor. *Could this really be it?*

I checked my email again before going to bed. Chip was burning the midnight oil, too. He had already replied to my message.

> *It's an incredibly interesting idea that would be worth thinking about—although it might be slightly ahead of its time... If you can find some phages with activity against* Acinetobacter, *I will give the FDA a call to see if they will issue an eIND for compassionate use.*

Chip's positive response should have been nothing but exciting. But at first, I could only focus on two words: "compassionate use." I stared at those words for a full minute. So, there we had it. Even Chip was admitting now that Tom was dying.

I could just hear him reviewing Tom's situation with Connie through his most objective lens. Chip was deeply empathetic, but he'd told me once how, as a physician, you have to compartmentalize. You can't let yourself get so emotionally wound up that you can't function as your patient's physician, or it means they don't have a physician, and that's worse. You have to be able to make scientifically and medically sound, pragmatic decisions. And the truth was, despite everyone's hard work and hope, there was no way you could convince yourself that Tom was getting better. He was barely communicating. His kidneys were barely holding on. He needed pressors to maintain a livable pulse and he needed the vent to get enough oxygen. And the superbug wasn't the only thing killing him. His underlying issues—the pancreatitis foremost—and the collateral damage meant that his body was breaking down, bit by bit. His organ systems were failing.

Our best hope now was that the FDA would decide that because he was going to die anyway, an experimental therapy was worth the risk.

It had been a long-running joke between Tom and me that wherever we traveled, Tom "collected" local parasites or weird infections, always returning home with some malady or another. Like Cameron and his Pokémon cards. Tom had even quipped once, after our MRSA experience, that it was his goal to collect all six of those deadly ESKAPE pathogens. It had seemed funny at the time: "Gotta catch 'em all!" Now his next acquisition—after *Acinetobacter*—could be an entirely new character from a new deck of cards, one with protective powers, maybe the ace up his sleeve.

At the hospital the next morning, I strode through the atrium feeling bathed in the light and the energy of possibility. The ride up to the TICU wasn't the usual discouraging descent into fear. And when I reached Bed 11, I was ready for a painful but important conversation. Tom and I needed to have another "life or death" talk. The first talk we'd had like this was in the Frankfurt ICU, more than two months before—*two months of this "near-death" experience.* Whatever I'd said that day had triggered his fight response. Granted, today the trach vent made it a one-way conversation, but we had to do our best.

I leaned close and took his hand in mine, hating the gloves I had to wear. I thought I detected Tom's lips move at my touch, which was a good sign. Maybe he was just conscious enough to hear me.

I told him the truth. The docs were now out of ammunition. They were all out of antibiotics, and he was not a candidate for surgery. So, if he wanted to live, he'd need to fight again. This would be a fight for time, while I looked for an alternative treatment I didn't know if I could find. And a fight against the continuing deterioration of his body, beginning to trigger organ shutdowns. No guarantees of anything, except certain death if we both stopped trying.

"Remember we had that talk back in the Frankfurt ICU, where I told you that if you want to live, you have to fight?" I began, and my voice faltered, catching in my throat. I swallowed and tried again. "Honey, I know you have been fighting so hard, and you're very tired. The doctors here are doing all they can, but they tell me they can't do anything else."

I knew that he knew this. In the still pause, I watched as a single tear welled up in the corner of his eye and spilled across his eyelashes. He blinked, but his eyes stayed closed as another tear trailed down his cheek. I let go of his hand, and wiped his face with a cloth. Later, I realized that I'd clenched my other fist when I saw that my nails had a dug four red half-crescent moons into the palm of my hand.

"I want to grow old with you, Tom. But I don't want you to live just because *I* want you to. That would be too selfish. This is your life, not mine." I took a deep breath. "And the thing is, it's okay if you don't want to fight anymore."

No clear response, if I was being objective, which admittedly was getting harder. He couldn't see me, but I was pretty sure he heard how my voice wobbled. I held his hand again gently.

"But if you want to fight, I'm going to fight, too. We're in this together. I will leave no stone unturned. In fact, I've been reading some articles on experimental treatment for multi-drug-resistant infections, and I have an idea..."

I told him about the phages, how they'd evolved over millennia to become the perfect predator of their hosts—bacteria. Of course, if he were awake, he'd be chomping at the bit, asking a million questions. But I had to cover all the bases, an effort at informed consent, in case he could hear me. So, I laid it out. Longshot as treatments go. Sound science as far as it went—but untested on humans infected with a fully antibiotic-resistant *Acinetobacter baumannii* strain that had totally colonized the body. Experimental, which meant it could take time to get permission to use on him. And no guarantees that it would work or, even if it did, that he'd recover from the damage already done.

"I'm not sure how to do it yet, but maybe we can get you some experimental phage therapy."

I squeezed his hand gently. "If you want to try it, can you squeeze my hand?"

Steff and Tom on their wedding day in Hawaii (July 26, 2004). *Courtesy of authors*

Steff crawling backward into the Red Pyramid, in Dahshur, Egypt (November 25, 2015). *Courtesy of authors*

Tom descending into the Red Pyramid, in Dahshur, Egypt (November 25, 2015). *Courtesy of authors*

Steff and Tom tour Luxor, Egypt, in a horse-drawn carriage the day he became ill (November 28, 2015). *Courtesy of authors*

The MS *Mayfair* cruise ship, docked in Luxor, Egypt (November 28, 2015). *Courtesy of authors*

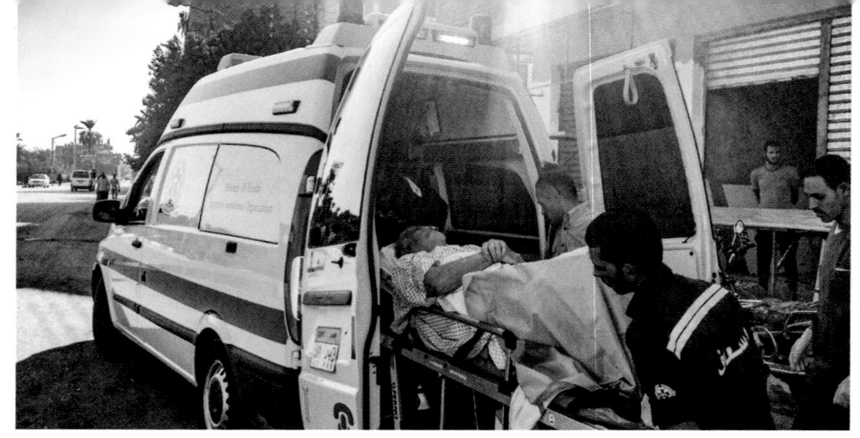

Tom on a gurney, being loaded into an ambulance outside the clinic in Luxor, Egypt, en route to the CT clinic (November 30, 2015). *Courtesy of authors*

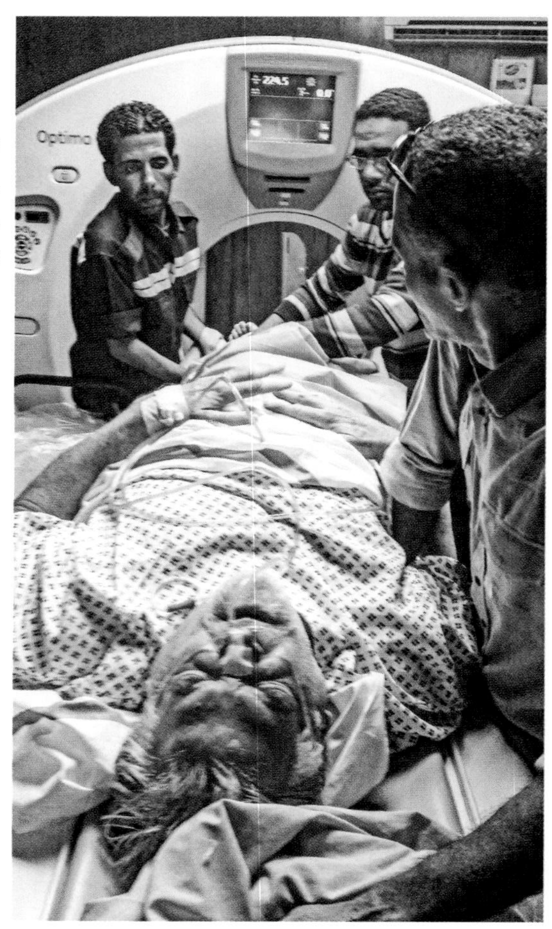

Tom in agonizing pain before receiving a CT scan in Luxor, Egypt (November 30, 2015). *Courtesy of authors*

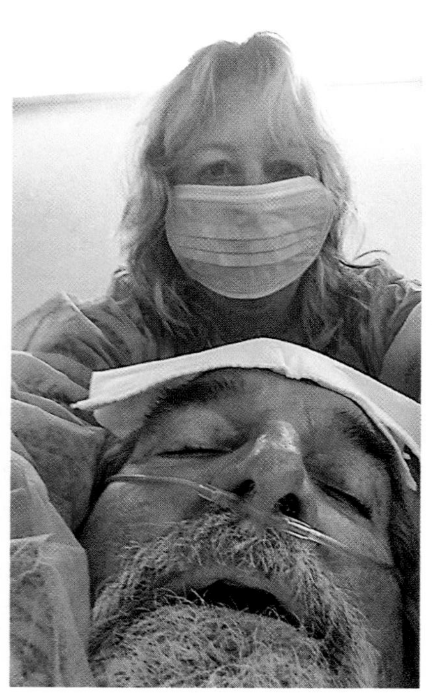

Steff takes a photo with Tom while he is fading from consciousness in the ICU at the Uniklinik in Frankfurt, Germany (December 6, 2015). *Courtesy of authors*

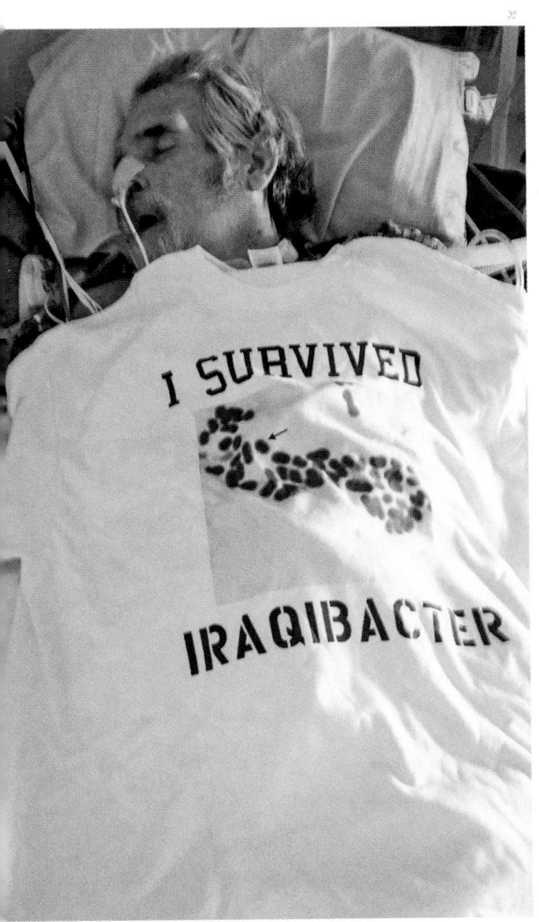

Tom in a coma, draped in the T-shirt emblazoned with I SURVIVED IRAQIBACTER on his sixty-ninth birthday (February 18, 2016). *Courtesy of authors*

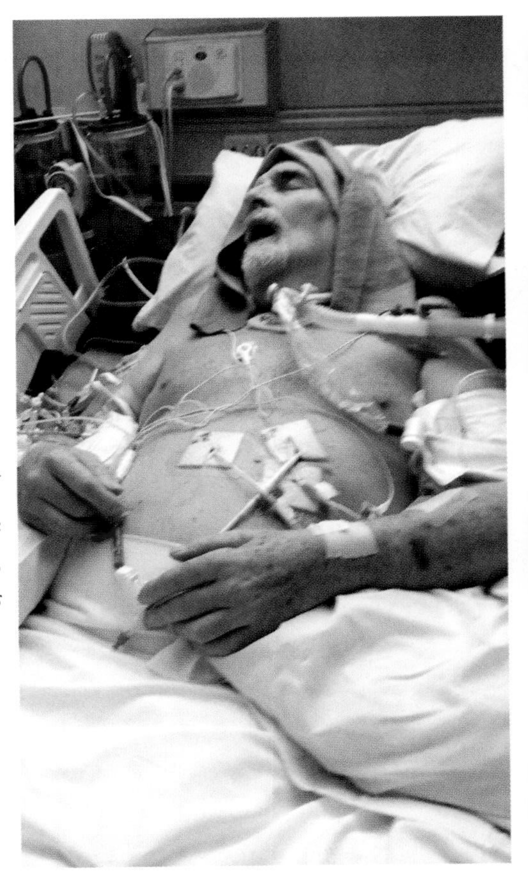

Tom in a deep coma, entering multisystem organ failure in the Thornton ICU in La Jolla, California, the day before phage therapy begins (March 14, 2016). *Courtesy of authors*

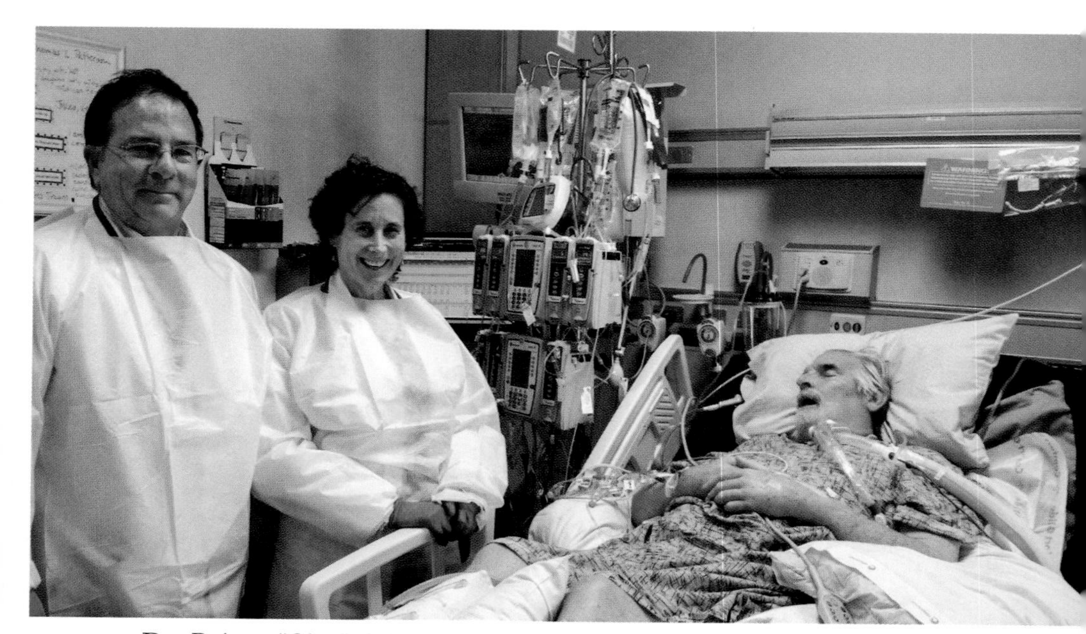

Dr. Robert "Chip" Schooley and Dr. Randy Taplitz prepare to administer phage therapy to Tom for the first time in the Thornton ICU (March 15, 2016). *Courtesy of Carly Patterson DeMento*

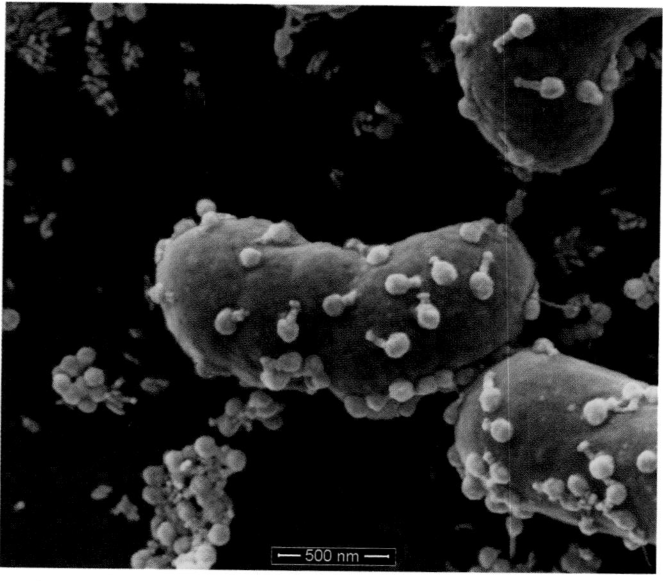

Scanning electron micrograph (magnification x100,000) of an *Acinetobacter baumannii* bacterium (stained blue) sampled from Tom, being attacked by some of the phages (stained green) that rescued him. *Credit: Dr. Robert Pope, National Biodefense Analysis & Countermeasures Center*

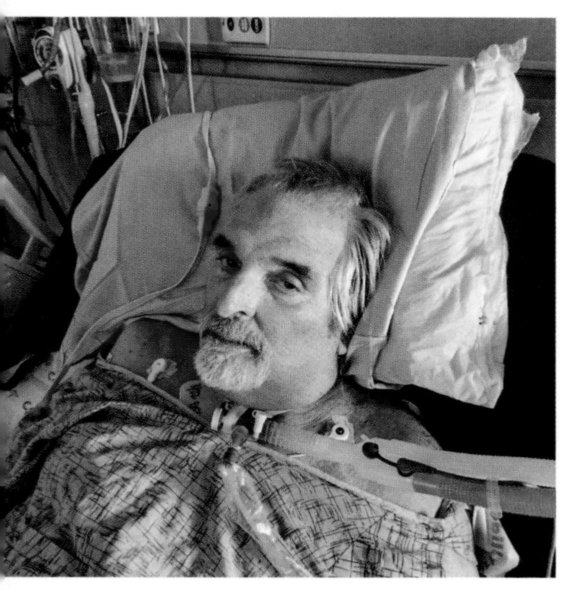

Tom wakens from his coma following a septic shock episode seven days after phage therapy begins (March 22, 2016). *Courtesy of authors*

Tom in a wheelchair getting some fresh air with his daughter, Carly Patterson DeMento, outside Thornton Hospital (May 16, 2016). *Courtesy of Carly Patterson DeMento*

Tom with Steff at their home in Carlsbad, California, two weeks after being discharged from hospital (August 28, 2016). *Courtesy of authors*

Tom's long road to recovery: working out in the gym (January 2017). *Courtesy of authors*

Tom holding a photo of *Acinetobacter baumannii* beside Steff, holding an image of a bacteriophage (May 2017). *Courtesy of UC San Diego Health*

He seemed to stiffen but... nothing. And then—he squeezed back, hard. *No retreat.*

That night, when I went to bed, even looking at the worst-case scenario with clarity, I didn't cry myself to sleep. I dreamed of wading waist-deep in a swamp, hunting for phages as if panning for gold. The water was murky and putrid, swirling with images of alien-looking phages, their heads the shape of microscopic geodesic domes and rocket ship tails trailing long filamentous fibers. When I looked down, I saw that I wasn't holding a miner's pan, but the cracked bedpan from the clinic in Luxor. I woke in a panic and rubbed my face with sweaty palms. With relief, I realized it was just a dream. But for the first time waking up to the real-life nightmare of Tom's illness, I felt more exhilarated, less hopeless. I leapt out of bed so fast I startled Newt and the kittens, who had curled around my knees while I slept.

Now all I needed was to find some phages. How hard could that be?

Tom: Interlude V

The curtain goes up on a play. I am in the audience, watching the actors passively. The room is a dull white, which makes it feel even colder and more sterile under the stark artificial light. Two sides of the room are almost all glass; on one side, people peer through a curtain of sphagnum moss into my room, the terrarium. There is a bed in the middle surrounded by a peat bog; the water, barely visible, is black, brackish. Each time someone steps into the room, a wafting smell of decay fills the air; it is the inevitable smell of death. An incandescent bulb on the ceiling flickers and buzzes next to me, like a blowfly waiting for carrion.

Beside the bed, a TV monitor flashes numbers and squiggles, and an alarm rings out. I cannot feel my body, which I know should frighten me, but I no longer care. I am floating above the bed, and I glimpse the top of an IV pole, which has five bags hanging on it, one half full of dark red blood.

Steff is in the play, too. She was nodding off in the corner offstage, but now she jumps up at the sound of the alarm and enters stage left. Rushing over to the bed, she pushes the call button with her gloved finger. Two actors suddenly appear stage right, a man and a woman with long, flowing robes. In seconds, they each pull yellow gowns over their heads and snap on gloves. They too, approach the bed.

Just then, I am struck by the sight of a snake lying in the middle of my bed, all curled up, immobile. Why I didn't see it before, I don't know. This confuses me. With a growing sense of dread, I know that I am allowed to stay alive as long as the snake is alive. But the snake is barely alive now. Its eyes are like slits. A web of green veins is visible on the surface of its eyelids, as if spun by a spider. The color of its skin is mostly jaundiced, its belly mottled with gray, black, and red bruises where it has been poked and prodded relentlessly. Its scaly skin is so thin, it is translucent. Through it, I can see that the only spark of life left is an ember in its tail; its glow ebbs and flows faintly, with the beating of its heart. Steff kisses its lips tenderly; they are bluish and covered with caked blood and spittle.

The snake licks its lips, and they are my lips. I am the snake. And I am dying.

The snake is being swallowed by a demon, which emerges like a volcanic island from the quagmire of the netherworld. The demon has a head like a lion and the jaws of a crocodile, the eater of souls. The demon unhinges its jaws, and I retch at its smell: vomit. I feel its teeth tear at my skin, which sloughs off in long papery ribbons and falls into the bog, where it will be absorbed by humification into gelatinous peat. As its asynchronous ratcheting devours me, accordion-like, my lungs gulp for air. I swallow black bile. A wave of gastric acid washes over my exposed heart, which is pale gray and barely pulsing. Is this the demon that has eaten Ra, the Egyptian sun god, when he disappeared into the underworld? If so, I will never see the sun again. I feel an overwhelming sense of despair; I have never felt more alone.

The man and woman circle the snake on the bed; the man inserts a bronchoscope. Its silvery segments slither into the snake's throat, and within seconds it sucks out a glistening plug of mucus. Oxygen restored, the cobwebs clear briefly from my eyes, and I am back in my body.

A group of three doctors walks toward my terrarium, their white coats flapping. I point at them vigorously. Steff laughs; she knows I am urging her to join rounds, where they will discuss my case like the specimen that I am.

As the brown curtain in front of my eyes closes, I think of phages.

Yeah, bring those little fuckers on.

15

THE PERFECT PREDATOR

February 21, 2016

What did it even mean to go on a phage hunt? Encouraged by Chip's support for attempting phage therapy, it had all seemed so possible yesterday, and even this morning. But after draining two cups of coffee, I still hadn't shaken the frantic phage hunt from my dream. I had already pored through a stockpile of research papers on PubMed and learned that there were an estimated 10^{31} phages on the planet—ten million trillion trillion—in existence at any given time, and no one really knew how many existed that preferentially attacked each species of bacteria. It could be dozens, hundreds, thousands. How would I ever figure out which ones might kill Tom's bacteria, and then find them in time? Identifying which phages might be effective against Tom's *Acinetobacter* made finding a needle in a haystack sound easier. I was suddenly overcome with the impossibility of the task I had taken on for myself. *Biting off more than you can chew*, my mother would say. She had always been my sharpest critic. I tried to shrug off my doubt. Tom was counting on me. So were the girls. I imagined what he would say if he could talk. "The enemy of my enemy is my friend," if he was feeling philosophical. "Hey, another killer bug to add to my collection," if his sense of humor was intact. He'd always been a risk-taker. So, there was no turning back now. Where to turn *to* was the question.

Who in the world is doing this work? Where do I even start? It'd been a long time since I'd felt like I was starting at square one like this.

Since phage therapy was outside the accepted repertoire of Western medicine, it was going to take outside-the-box thinking to figure out if and how it could be used now. What's more, that process sometimes takes decades or longer in the methodical world of science and medicine, and there just wasn't time for that. This had to be on a fast track, or Tom would die waiting.

As an infectious disease epidemiologist, I routinely track threads of data and discoveries across time, space, and populations to find fresh insights. Now, as I read about the discovery of phages and the first attempts at phage therapy one hundred years ago, it was beyond ironic that Chip had mused that phage therapy was an idea ahead of its time. Thanks to the internet and our online university library access, I was reaching back in history to published articles that detailed the basic science that scientists had stumbled upon in the days when Model Ts ruled the road. The authors of these studies—my new mentors—had died long before my PhD adviser had.

Articles about phages ranged from clinical and basic science to studies of the evolutionary path they are believed to have traveled since emerging with the first terrestrial animals some 450 million years ago.

Most scholars now credited the "discovery" of bacteriophages to Félix d'Hérelle, a scientist who observed "filterable agents" that killed bacteria in 1917. But clarifying how this antiseptic agent killed bacteria would turn out to be a bigger challenge. In 1915, an English bacteriologist named Frederick Twort was working to develop a smallpox vaccine, but his lab cultures were often contaminated with *Staph* bacteria. Upon further examination, he observed small, glossy spots in the thin film of *Staph* in the Petri dishes where the bacteria didn't grow—just as we'd seen in my virology lab class in 1986. Twort showed that whatever had killed the *Staph* could pass through a super-fine porcelain Pasteur filter and infect new bacterial cultures. Pasteur filters, so named for the acclaimed father of microbiology Louis Pasteur, kept larger microbes like bacteria from passing through. So, whatever this agent was must be smaller than bacteria. Twort was unsure just what this was. But his findings—the discovery of a "bacteriolytic agent" that killed bacteria—were published in the top medical journal *The Lancet.*

Two years later, Félix d'Hérelle performed similar experiments but took these observations a step further. He was convinced that these bacteria-killing agents were some kind of new life form, which he proposed was a virus. He and his wife named them "bacteriophages." There was fierce debate at the time about whether Twort or d'Hérelle had discovered phages first, and whether they were viruses or enzymes, because no one at the time knew what phages looked like, just that they seemed to reliably destroy bacteria.

The more I learned about Félix from the writings of his biographer, medical historian Dr. William Summers from Yale, the more he felt like a kindred spirit. He grew up in Montreal, so he considered himself Canadian, like me. He had been ostracized by his peers and was a bit of an oddball. A Google search turned up another coincidence: the University of Toronto, my alma mater, had published one of his monographs on phages in 1922. With a click of a button, I ordered a copy of the English translation from Amazon. It quickly became my bedtime companion.

Although Félix was ridiculed as a "vagabond scholar" with no formal education in science, as one of the earliest applied microbiologists, his "microbe-centered worldview" had been noted for its prescience. Attempting to model his career after Louis Pasteur, who pioneered our understanding of the role of microbes in alcoholic fermentation, Félix tried unsuccessfully to develop maple syrup into whiskey.

He'd even traveled to some of the same regions in his scientific explorations as Tom and I had. In 1907, he was hired by the Mexican government to continue work on fermentation. By 1909, he had succeeded in turning a species of agave into schnapps—which probably tasted similar to the age-old *pulque* "agave wine" that Tom and I had sampled in our travels there.

Félix also had a natural-born curiosity. In 1910, when a plague of locusts swarmed the Yucatán, the local natives showed him a place where the locusts were dying from an unknown disease. Félix noticed that these dead locusts were surrounded by copious amounts of black diarrhea. Culturing the locust poop, he concluded that they had died of the insect version of blood poisoning related to a *coccobacilli* bacterial infection. He seized on this finding, showing that by dusting *coccobacillus* cultures on the crops locusts devoured, without apparent harm to humans, he could wipe out the locust infestation. After using this approach to combat locust plagues elsewhere in South America and North Africa, he became famous as the "father of biological pest control," but his scientific contributions didn't stop there.

Félix astutely observed that some locusts seemed impervious to the *coccobacillus* infection, and when he streaked the *coccobacilli* onto agar with the poop from the locusts that recovered, he observed that clear spots formed around some of the bacterial colonies. Something was killing the bacteria, but what could it be?

It wouldn't be until he moved to Paris years later that he would solve the mystery, and it was better than any *Forensic Files* episode. During the heart of World War I, Félix was working at the Pasteur Institute when an outbreak of dysentery occurred. Asked to assist with the outbreak investigation, he inoculated stool samples from the patients who survived into culture medium and let them incubate for eighteen hours, then filtered them. He then added the filtrate to test tubes that contained a culture of the *Shigella* bacteria causing the dysentery. The test tubes were initially cloudy, a virtual bacterial soup, but the next day, they were completely clear! He looked at the solution under a light microscope—no bacteria. That's when he had his eureka moment. When he discovered that the bacteria-killing agent was able to pass through the Pasteur filter, he concluded that the agents responsible were smaller than bacteria and must be a virus that preyed upon them.

How could you not admire Félix, who in 1917 would write about that moment in the lab this way:

> ...on opening the incubator I experienced one of those rare moments of intense emotion which reward the research worker for all his pains: at first glance I saw that the broth culture, which the night before had been very turbid, was perfectly clear: all the bacteria had vanished...as for my agar spread it was devoid of all growth and what caused my emotion was that in a flash I understood: what causes my spots was in fact an invisible microbe, a filterable virus, but a virus parasitic on bacteria. Another thought came to me also. If this is true, the same thing will have probably occurred in the sick man. In his intestine, as in my test-tube, the dysentery bacilli will have dissolved away under the action of their parasite. He should now be cured.

"Rare moments of intense emotion" at the sight of the spotty Petri dish? His detached description of emotions made me laugh out loud. If Tom were here, he would say that Félix was speaking my language, and we would have laughed over it together. *We will someday*, I told myself, but that would have to wait. Tom might find a kindred spirit in Félix, too, in that they both were inclined to self-experimentation to try out new ideas in treatments. Tom had once made a disastrous attempt to desensitize himself to poison oak by ingesting it over the course of several days. (He got the worst case ever. Don't try this at home.) Félix, determined to help sick children in an outbreak of dysentery, had pushed ahead on his own. Since phage therapy had not yet been used to treat a human bacterial infection, he first tried his phage prep on himself. Sounds crazy, but self-experimentation was actually pretty common a century ago.

He then gave a weakened version to a twelve-year-old boy who had severe dysentery, and his symptoms disappeared after a *single dose*. Three additional children were treated, and they started to recover within twenty-four hours, too.

With only a rudimentary knowledge of phage biology, I spent a few hours that morning scouring PubMed to brush up. A century after Félix discovered bacteriophages, scientists now understood a lot more about them. A single drop of water can harbor a trillion phages and they are found virtually everywhere—in soil, oceans, and our bodies. The way phage hijack bacteria and use them to replicate was straight out of science fiction—*Invasion of the Body Snatchers*—only it wasn't fiction. No wonder phages are the most numerous creatures on earth. They are astoundingly efficient microscopic machines. Connect. Inject. Replicate and release, destroying the bacterium's cell wall in the process.

Despite the phage's impressive takedown of the host bacteria once inside them, phages do face a few challenges. First, they must evade the body's immune system and defensive responses. Second, phages must outrace bacterial resistance, which can occur in a matter of minutes, as bacteria that are susceptible to a particular phage die, and mutant bacteria that are resistant to the same phage flourish, with new space to take over. In the same way bacteria have developed adaptive strategies to resist antibiotics, they use a host of tactics to fend off phages, including their own kind of immune systems, called CRISPRs. Bacteria may block, disable, or "change the locks" on the receptors the phages need to gain access. They may modify molecules to toughen their slimy protective shield, or create internal mechanisms that interrupt the phage's replicating process or interfere with the virions' ability to assemble for the final destruction of the host cell.

But two can play at that game, and do. Phages evolved to prey on bacteria, and although bacteria had gene-slicing CRISPR mechanisms to defend against at least some of them, phage have evolved anti-CRISPR defense mechanisms so they can continue the assault. Moreover, there are billions of different phages, many having evolved to target the same bacteria but through different receptors. The task for researchers is to find the ones that match up with a target bacterium, purify them so they're safe to use, and then deliver them into the body—in this case, Tom's.

The evidence was persuasive. On paper, phage therapy appeared to be ideal for treating multi-drug-resistant bacterial infections like Tom's. I ticked off their advantages on my fingers. Phages are the natural predators of bacteria, and many of them can be found for targeting the harmful ones without hurting the "good" ones. Since phages are plentiful in nature, theoretically they should be relatively easy to find for targeting many bacterial pathogens. They dwindle in number when their host bacteria are sparse, and grow when and where that bacterial population is plentiful, as in an infection. When the host bacteria thin out, phages adrift are filtered out by the liver and spleen, where specialized cells in the immune system engulf them, then digest them, and they're gone. This means that they largely disappear once their job is done. Antibiotics exert much longer-term effects, with known side effects that can damage human tissues and upset the natural balance of your microbiome. But would phages work in practice the way they work in theory? The jury was still out on that.

It wasn't that nobody had tried to find out. Phages and penicillin were discovered in the same era—phages in 1917, penicillin in 1929. But why was a successful treatment discovered one hundred years ago not being used more widely? The surprising lag time was a factor that had echoed through the history of both phage therapy and penicillin. The miracle of penicillin had languished for a decade between its discovery and commercialization by pharmaceutical houses, because it took time to isolate, purify, and expand it for the masses. The growing casualties of World War II created an imperative that had finally prompted action.

But a hundred-year delay? After his discovery, Félix had gone on to apply phage therapy to thousands of cases of cholera and bubonic plague in India. His big ego and tendency to show off drew international attention. He attracted an ever-widening circle of followers and became Sinclair Lewis's inspiration for his 1925 Pulitzer Prize–winning novel, *Arrowsmith*. All the hubbub increased the popularity of phage therapy for a while.

But there were critics and complications. Although Félix insisted on matching his phage preps to a patient's individual infection, that wasn't considered practical, so some companies just manufactured the phage preparations in the lab for general use. Eli Lilly, a division of Abbott and a company that was later acquired by L'Oréal, started selling phage preparations to treat wounds and upper respiratory tract infections. Some had trendy names like Staphylo-gel. Soon there were problems. A few were sold with exaggerated claims that they could cure viral infections—like herpes—which they couldn't possibly do, because bacteriophages target bacteria, not other viruses. And until the late 1930s, phage cocktails were often sold without being purified, so some actually could have caused harm. Other manufacturers mistakenly sterilized their phages with agents that were supposed to "stabilize" them. All told, many phage preparations being sold were useless.

Within the scientific community, phage therapy faced major challenges from the get-go. For roughly thirty years after Félix identified them, scientists still didn't have the technology to see them. Some scientists, including some Nobel laureates, pooh-poohed Félix's claim that phages were microorganisms; they believed phages to be enzymes. It wasn't until the first electron microscope was developed in 1940 that bacteriophages could finally be visualized, and Félix was vindicated. Today, new treatments undergo randomized clinical trials and bioethics reviews, but these standards and review processes weren't established back then. Without proper controls and safety checks, phage therapy couldn't be trusted.

After penicillin came to market, phage therapy was relegated to the back burner, at least in North America. But it was more than science that cooled commercial interest. There were political reasons, too. I was struck by the fact that one of the few centers offering it was in Tbilisi. The story of how that came to be was as surprising as the phages themselves. After World War I ended, Félix met a young Georgian bacteriologist named Giorgi Eliava at the Pasteur Institute in Paris. They hit it off, and Eliava worked with him on some phage projects. Eliava returned to Tbilisi in 1923 to resume his position as head of their modest microbiology institute. His dream was for it to become the first center in the Soviet Union to feature phage therapy. He pulled it off in 1926.

The new center was awesome, situated on the bank of the river that runs through Tbilisi, amid a cypress grove. It even had Stalin's blessing, but that connection gets stranger. In 1934, Eliava invited Félix to visit and help make their center the world's premier phage therapy institute. Félix spent six months in Tbilisi. After Félix planned to relocate there permanently, a home—"d'Hérelle's cottage"—was erected and dedicated to him on the grounds of the institute.

But if there was a bromance between our phage pioneer and Stalin, it was not to last. Eliava was arrested and declared an "enemy of the people" by his nemesis, Lavrenti Beria, the chief of secret police, whom Stalin referred to as "our Himmler." Beria saw to it that Eliava was executed in 1937; he was one of many scientists to suffer a similar fate around that time. Devastated by Eliava's death, Félix never followed through on plans to return, so he never lived in the cottage constructed in his honor. In fact, it became occupied temporarily by what would eventually become the KGB, who later came under Beria's command.

What later became known as the Eliava Phage Center struggled initially but became renowned throughout the region and internationally. At its peak, in the 1980s, it employed about eight hundred people, including more than one hundred phage researchers. They manufactured phage cocktails, sprays, salves, ointments, and tablets—up to several tons a day—about 80 percent of it going to the Soviet army, mainly to treat dysentery.

In 2016, the main phage therapy centers were the Eliava; a second one in Tbilisi, and another one in Wrocław, Poland. They had decades of experience with phage therapy, and despite the lack of endorsement by Western medicine for so long, people still flew there regularly from the West to get different kinds of infections treated.

Why didn't phage therapy get replaced by antibiotics in Eastern Europe like it did here? Since the discovery of penicillin was considered a military secret in the early years of World War II, at first, they didn't even know about it. And even after they did, penicillin wasn't available there on a consistent basis. There, as in the United States, penicillin was so hard to manufacture in sufficient quantity that until production techniques were perfected, it was so precious that it was recovered from the urine of patients who received it. No such problem with phages. They turned out to be ideal for treating battle wounds. The Japanese military used phages, and so did the Germans; some were found in medical kits used by Rommel's forces in North Africa. And the Russians used phages to treat wounds in their war against the Finns during World War II, and more recently in Chechnya.

Apart from the logistical issues, another reason why phage therapy didn't get taken up in the West is because if you supported it, you would be labeled a pinko commie sympathizer. This "Russian taint," wrote Dr. Summers, was enough to scare off most Western scientists, research funding, and commercial interests alike. People really started to jump off the bandwagon—especially in the US—when the American Medical Association published a series of damning reports in the 1930s and '40s concluding that there was limited data to support the efficacy and reliability of phage therapy, except for its use in treating staph infections. The hardened position against further study to advance phage science reeked of political bias not only geopolitically, because of its popularity in Russia, but within the Western scientific community itself. Isolationism is no friend of true science.

One of the few researchers who kept popping up in the older phage literature was the now retired NIH biologist Carl Merril. Nearly eighty, he had worked to advance phage therapy in the US for fifty years—longer than I had been alive. But the politics of that era had shuttered interest in phage research here, and the limitations of technology and the incomplete data available from countries where it was used meant that Carl faced an uphill battle.

I would learn later that despite the obstacles Carl faced in his 1970s experiments at his NIH lab, he had found that when phages are systemically administered to lab animals, they are destroyed by the liver and spleen. In the mid-1990s, he and his protégé, Biswajit Biswas, learned to select phage strains in the lab that could evade the filtering action of these organs so they could remain in circulation longer and thus serve as more effective antibacterial agents. By 2002, they used phage therapy to successfully treat mice that were infected with the superbug *Enterococcus faecium* that were resistant to vancomycin—one of the big-gun antibiotics. Following these productive animal experiments, Carl tried to get the NIH to support clinical studies but had no success. Eventually, trends in funding and internal politics turned against him completely. Pressured to retire, he finally did, but he never abandoned his scientific conviction that phage therapy, further refined, could save lives.

I couldn't get over the irony. Just as phage therapy was abandoned in the West, phage biology—the pure science of it—was taking off in other ways. If you look at the early Nobel Prize laureates in basic science at the time, about half of them were phage researchers. Much of the work had been done in the 1940s and '50s, receiving recognition only much later. Phages had been used to show how genes turn on and off. Phage enzymes had launched the fields of molecular biology, genetic engineering, and cancer biology. More recently, phages were critical in seminal research by Drs. Jennifer Doudna and Emmanuelle Charpentier that identified the CRISPR-cas9 gene editing system that is revolutionizing synthetic biology.

Félix had died long ago, and Carl had been put out to pasture. There seemed to be relatively few phage researchers left in North America who believed in its therapeutic potential. If there were any working on phages active against *A. baumannii*, I had to find them. Now that we had the science and the technology to see and study phages, tweak their genes, and tailor their appetites, maybe it was time for their comeback.

••••••••

On the way to the hospital a few hours later, I called Carly and then Frances to tell them what I had in mind. How do you tell your stepdaughters that you want to inject live viruses into their dad to try to cure him of his bacterial infection? It sounded ludicrous, but I gave it my best shot. Both had listened attentively and were encouraged that Chip was supportive.

"So, basically, this is like a green alternative to antibiotics, right?" Carly asked when I reached her at home. Carly lived about ten hours away, in a Victorian farmhouse just north of San Francisco, where she and Danny had moved shortly before their wedding. I could imagine the pastoral setting with song sparrows trilling in the background. Tom would have identified them—*Meliospeza melodia*—if he were here. If he could talk.

"I never thought about it that way, but yeah, I guess you could say that," I replied. I was on my speakerphone, sitting in bumper-to-bumper traffic on I-5. "And over millions of years phages have co-evolved with bacteria, even those that have become superbugs."

"Love that Mother Nature," said Carly wryly. "But where are you going to find the phages?"

"That's the six-million-dollar-question. Haven't figured that part out yet," I admitted.

"But where there's a will, there's a way."

I was relieved she was on board. One down, one to go. I reached Frances while she was out doing fieldwork under contract for a biological sciences project.

"Have you watched too many reruns of *Forensic Files*, Steff?" She was teasing me. I knew she detested the show. In an episode that I'd forced her to watch with me when she had visited last, a scientist had used swamp plankton to solve a murder that was decades old.

"At least I am not a black widow trying to finish off her husband with succinylcholine," I countered, referring to what often turned out to be the murderer's preferred poison. "Besides, your dad's illness belongs on another show: *Monsters Inside Me*."

After I explained the idea behind phage therapy, she gave it her blessing: "That there is some cool shit."

Truer words had never been spoken. Since phages are plentiful in the guts of humans and animals, billions of them are excreted into the environment every day in fecal waste. So the first places to start looking for Tom's cure would be locations with a high concentration of that particular organic matter: feces. This meant that one of the best places to find them in was in raw sewage. In other words: shit. No, I thought, that detail could be explained later, if and when the time came. I could see the *National Enquirer* headline now: MAN CURED OF DREADED SUPERBUG WITH PURIFIED SEWAGE.

16

SEMPER FORTIS: ALWAYS FAITHFUL, ALWAYS STRONG

February 21–26, 2016

The evening of February 21, intent on finding a clue that could lead to phage sources, I spent the night with PubMed and half a bottle of 7 Deadly Zins. Newtie and the kittens made a nest along my side in the blanket that covered my lap, each jockeying for a position that would reward them with a scratch under their chin or on the belly. By eleven p.m., I had finished scribbling down a list of the top phage researchers who made any mention of *Acinetobacter baumannii* in their published papers. I highlighted the ones that were based at US institutions, knowing that time was of the essence. The AppleTV connected to our surround-sound speakers was set on random play, and I clicked it on to fend off the painful quiet. The room filled with one of my favorite songs, "Courage," by the Tragically Hip. We were living the through line: sometimes all we can do is stay upright through the worst of times, make the best decisions we can, and live with the consequences, never knowing what they'll turn out to be.

My mind wandered. It had now been over a month since I had last heard Tom's voice. I took a deep breath and hit the answering machine button, just so I could hear his deep baritone: *Hi, you've reached Tom and Steff. We're not here right now...*

"OK," I announced to the kitties. "This is no time to feel sorry for myself." I thought of Tom fighting for his life against this superbug that I'd once dismissed as wimpy. He'd made the decision to fight, and we'd made our pact to do it together. As I sang along with the chorus to "Courage," I summoned whatever strength was left in me to get the job done.

My search had turned up only a handful of people in the US who could maybe—just maybe—help find the right *A. baumannii* phages for Tom, and were close enough to possibly beat the clock. I stared at the blank email page. It stared back. I would have to make cold calls, emailing someone out of the blue. I was used to that, since I often contacted leading researchers for their expertise. But this felt weird for a few reasons. First, although I had an undergraduate degree in microbiology, phage research was far afield from the discipline I knew best and for which I was known by others. Second, this wasn't a professional query. It was as personal as it gets. I was about to ask a group of strangers to help me save my husband's life. A proverbial shot in the dark, with Tom's life hanging in the balance. But if Tom died despite my efforts, at least I'd know that I'd tried my very best.

In fifteen minutes, I'd written a generic email that I tailored slightly for each of the researchers on my list. The first one was to Dr. Ryland Young at Texas A&M University, director of the Center for Phage Technology. He had been quoted in a news article in a top scientific magazine, where he claimed that finding phages to target a specific bacterium is "relatively easy." Could this possibly be true?

Dear Dr. Young,

I learned about your phage research in a commentary in Nature. *As an infectious disease epidemiologist, I find this approach to treating clinically resistant ESKAPE pathogens fascinating. However, I also have a personal interest, as my husband has a serious case of gallstone pancreatitis, which is complicated by a pan-resistant* Acinetobacter baumannii *infection that he probably acquired in Egypt, where he fell ill. After three months of acute illness, he is now being treated at a UCSD hospital by my colleagues, some of whom are top infectious disease experts. However, his condition is deteriorating because the infection cannot be controlled, and we are considering less conventional approaches. I recognize that your lab primarily does* in vitro *research, but I am wondering if you have any suggestions for us in terms of phage treatment. Many thanks for considering this unusual request.*

As I tailored each letter and hit the Send key, I could feel my desperation rising. Would he even read my email? Tom and I got so many phishing emails every day, junk disguised as urgent messages, or predatory journal editors offering to publish a research paper—for an astronomical fee. He and other researchers did, too, no doubt. Even if they opened my note, how could I expect a total stranger to respond to such an unusual request—and do it right away?

The next morning, I had a hundred or so emails waiting for me, as usual. Most were routine; others were junk. I scanned them briefly, triaging some of the work-related ones for later and deleting others. Several were from phage researchers I had written the night before who had sent regrets about my situation. A few explained politely that they had no *A. baumannii* phages. Others said that their approach was not yet ready to be tested in humans.

The last email I opened was from Ryland Young at Texas A&M. As I clicked open the message, I braced myself for yet another rejection. Like the other researchers, he wrote that he was sorry to hear about Tom's condition and the desperate circumstances, that Tom's case was heartbreaking. But unlike the others, he offered to help locate appropriate phages that might be able to treat Tom. He suggested we talk by phone ASAP and sent me three phone numbers where he could be reached. He signed it *Ry*. I was electrified.

I called Ry and we spoke for nearly two hours. It was as if I'd dropped like Alice down the rabbit hole. He proceeded to give me a crash course in phage biology, picking up where my virology class had left off thirty years ago.

I jotted down some notes outlining what we were up against. Unlike some phages, such as those active against MRSA, those that are active against *A. baumannii* are type-specific. This meant that we would need to match specific phages not just to Tom's bacteria's genus and species, but to Tom's actual bacterial isolate. In other words, it wasn't enough to know that they needed a phage match for *A. baumannii*. They would need to get a sample of the specific *A. baumannii* that was growing so successfully in Tom. Then they'd be able to turn to screening the phages they found to find a match. Did we have a sample of his isolate on hand so that any phages Ry or others found could be appropriately matched? We did. I would ask Sharon Reed in the UCSD microbiology lab to ship it to them right away.

Next, Ry asked me whether I happened to have any soil or water samples from Egypt, even soil from our hiking boots. If Tom's superbug was acquired in Egypt, Ry explained, environmental samples from the region would be potential sources for identifying phages, because in nature, they coexist. I recalled how dry the earth was when we visited Saqqara and Dahshur, and doubted that we would have any dust on our boots that would meet his needs. We didn't. No help there.

Last, Ry explained that we needed to find not just one phage active against Tom's isolate but, ideally, several. "As you know, bacteria have an uncanny ability to mutate," he explained. "Even if we are lucky enough to find a phage that we know matches Tom's isolate—and I have to warn you, we will be extremely fortunate if we do—if we treat him with a preparation based solely on this one phage, his bacteria will likely become resistant to it, almost immediately. To make it perfectly clear what we are up against, our lab has only collected a few *A. baumannii* phages in the last seven years. But I'll have them tested against Tom's bacterial culture as soon as we get his sample, and we will know if we have a match in a day or two. I'll also ask my team to plate out his bacterial culture against our environmental samples. You know what that means, right?"

I'd done my homework.

"Sewage," I replied.

"Yeah, these samples come from sewage runoff, swamp water, farms—basically anywhere you can find poop," he said.

"How many phages do we need to find, then?" I asked, with a growing sense of alarm.

"No one really knows," he replied. "Phage therapy is virgin territory in North America. The Georgians and the Poles both have centers, so they have the most hands-on experience. But they haven't collected enough empirical data to convince the FDA and other regulatory bodies that phage therapy can work. And most of their cases are more garden variety: *Staph*, *Pseudomonas*, and *Klebsiella*. Iraqibacter has become so virulent and resistant, it stands out, even among superbugs. My guess is that we'd need even more phages to cure a patient like Tom, who is fully colonized. Let's aim for three or four that we can grow here and send back to you in a phage cocktail."

"That would be amazing," I told him, my voice wavering. "I am so touched that you are willing to go out on a limb to help a total stranger."

"I'm the same age as your husband," Ry replied. "An old fart. About to retire. Maybe that's why his story struck a chord. And I'd like to see some real-world good come from this research and the career I've spent on phage biology."

Self-effacing humor aside, there was nothing retiring about Ry's style, and his old-fart status only meant that he knew a helluva lot about how phage therapy had developed—and stalled—through the years.

"There are a handful of researchers around the world who have kept plugging at the notion of a rebirth of phage therapy, but in many cases, they've been pushed to the margins of the phage community, some treated like pariahs," Ry said. "Remember, phages were discovered long before Watson, Crick, and Rosalind Franklin figured out the mystery of DNA's double helix, so the early research had its flaws. But I truly believe it's time for phage therapy to come of age, now that even the best antibiotics are becoming increasingly useless against superbugs. An effort like this, if successful, would give them credibility—and perhaps more funding for phage therapy research."

Ry brought a born-again fervor to this quest as an opportunity he saw to advance phage therapy research, as he'd spent much of his earlier career as a self-professed "anti–phage therapy" guy. As he told it, in his grad school days in the early 1970s, he'd been persuaded by the prevailing prejudice within the molecular biology community that regarded phage therapy as a "bizarre chapter" in medical history that should remain closed. That camp held that phages had been and still were very powerful as tools or models for discovering fundamental aspects of molecular biology, but their practical use in clinical applications—treating patients—amounted to quackery. Pipette dreams.

But, in 2002, at a national microbiology conference he'd been swayed by a very different vision of a modern approach to phage therapy, put forth by an internationally respected biotech scientist-entrepreneur from India. Thus enlightened, Ry became an advocate for a broader approach to phage science that included phage therapy. He eventually won the support of the higher-ups at Texas A&M, and in 2010 they established the Center for Phage Technology (CPT), providing faculty positions and a base research budget. It became one of the pioneering phage centers in the US. And it was this body of work by Ry and the CPT since, published in the ranks of influential peer-reviewed journals that had once deep-sixed phage therapy, that had caught my attention in my online search.

"If this works, it will be a game changer," Ry said.

"In any case, we have our work cut out for us," he said. "With your permission, later today I'll forward your email to everyone I know who might have some *A. baumannii* phages. With any luck, they'll have some they can send to my lab, where we'll test their activity against your husband's strain. Ship me his isolate *hasta pronto*. And email me any other information you think could pull at their heartstrings as you did mine. I'll set up a Google drive so we can share files."

I hung up, exhilarated. My heart felt like it was beating out of my chest—if I'd been hooked up to Tom's cardiac monitor, it would be sounding the alarm. I immediately sent Ry the photo of Tom I had taken with his Iraqibacter T-shirt, along with a photo of Tom from 2012 when he was in perfect health. I added Tom's academic bio and his curriculum vitae, which was over a hundred pages.

Chip was just getting to his office at the university when I caught him on the phone to share the good news.

"You won't believe this, but Ry Young from Texas A&M already responded to my email and offered to head up a phage hunt. So I guess I should ask you formally—if they find at least one phage, will you be the principal investigator of the protocol?"

This was a critical question. There were two reasons why I couldn't be PI of the protocol myself. First, I'm not an MD, and clinical protocols required one. Second, I was the wife of the patient, which meant I had a major conflict of interest.

I could hear a *whoosh* as Chip sat down in his office chair.

"Wow. That is quite a feat," he said. "And sure, of course I can be PI. I'll get hold of my contact at the FDA to get paperwork started for a compassionate use request. We need to have our ducks lined up so we're ready to go if they find a phage."

Chip wasn't surprised to learn that Ry needed to match phages to Tom's bacterial isolate and that they'd need to find more than one to overcome resistance. Chip chaired an NIH review panel that adjudicated research proposals focused on antimicrobial resistance, so he was also well aware of the regulatory challenges that lay ahead in the FDA's review process for an eIND—experimental investigational new drug. These are treatments of last resort, not yet FDA-approved and used only when someone is dying and all conventional treatments have failed. Tom qualified on both counts. And there couldn't be a more qualified clinician than Chip to shepherd this project. The question was, could the approval process move quickly enough to get any phage they happened to find sent in time to save Tom?

"This might be a stupid question," I said, "but why do we need an eIND from the FDA? Phages aren't medications."

"The FDA doesn't have any other mechanism for approving new therapeutics," Chip explained. "That's part of what discourages developers from testing new treatments like phage therapy. The FDA needs a new regulatory model, but the odds of that happening any time soon are nil. Though if we can cure Tom with phage therapy, that might be an impetus for them to make some changes."

The next morning, my inbox had more emails than usual. There had been a flurry of activity overnight, thanks to time zone differences. Ry's email had been persuasive. Phage researchers in India, Switzerland, and Belgium had all agreed to send *A. baumannii* phages to Ry for testing. Dr. Jean-Paul Pirnay, the head of the Belgian team from the Queen Astrid Military Hospital in Brussels who collaborated on the paper I had found by Dr. Merabishvili, wrote that his team had some phages active against *A. baumannii* that they were hoping to use topically to treat soldiers with burns. Incredibly, he offered to send their phages in a diplomatic pouch.

Chip had news to report, too. He had called his contact at the FDA, Dr. Cara Fiore, prepared to explain what phage therapy was and why it was needed for compassionate use in Tom's case. But Dr. Fiore, a microbiologist with the agency's Center for Biologics Evaluation and Research, knew all about it and was familiar with all the main labs in the US that were working with phages. She even offered to provide contact information for a couple of labs that weren't on our list.

This was so much better than either of us had even hoped for.

"Who heads up the other two labs?" I asked Chip. I was so sure I'd been thorough in my search. It seemed strange that I would have missed any that involved *A. baumannii*.

"Would you believe the US Army and Navy?"

I gave a low whistle. "You're kidding me. The military is working on phage research? No wonder I hadn't turned up any details on these investigators. Do you suppose it's classified?"

"Well, I don't know about that, but with all their service members coming back from the Middle East with multi-drug-resistant infections, it makes a lot of sense that they'd already be working on this," Chip said. "I have a call in to both later this afternoon."

The next day, my cell phone blasted Aretha Franklin's "Respect," the ring tone I had recently assigned to Chip. I picked up and could tell right away that his dander was up—his slight Southern drawl had a bit of an edge.

"I spoke to the brass at the Army and Navy who are heading up their phage programs. Both were leery of getting involved in the care of a civilian." He sighed, exasperated. "I told them that *I* would be the one overseeing the clinical protocol—that all they needed to do was to send the phage. They dug in their heels until I told them that I was more impressed with the Belgian military because they were offering *their* phages in a diplomatic pouch."

This was quintessential Chip. My nickname might have been Pit Bull, but Chip was like a bulldog guarding a bone when someone stood in the way of what was best for a patient. This trait had made him a few enemies, but more often it earned him deep respect from patients and professional peers alike. Under these circumstances, respect could be the trump card.

"Bottom line—the Army is definitely out," he said. "But I think I convinced the Navy to at least test their phage collection against Tom's isolate. If there turns out to be any antibacterial activity, Theron Hamilton, the lieutenant commander, said he can cross that bridge when we come to it. The approvals will need to go right up the chain to the admirals."

Theron would be putting a lot on the line to do this. But then, everything the Navy could learn about *A. baumannii* and how to treat it added to the Navy's medical arsenal if it struck the troops, and it could prove valuable in a bioterrorist attack. So, helping Tom could be a win-win. Tom would have been the last person to volunteer for military duty back in the sixties, during the Vietnam War. Now that the battle was right under his belt, it looked like he might get a chance to volunteer after all.

"Well, I still call that progress," I replied, holding the cell phone on my shoulder as I chased Newton away from the kittens' food. He was getting a pot belly.

"Theron warmed up to me after I told him that he described their 'Egyptian collection' like a sommelier describes a fine wine," Chip joked.

"He told me he was ordering his lab to work double shifts this weekend, so they'll test Tom's isolate against their phages as soon as we can ship them a sample."

So now we had not one lab but two working around the clock on the phage hunt, in the span of a few days. In fact, it was astonishing progress.

"I can hardly wait to tell Tom—even if he can't hear me."

We couldn't know yet if the labs would find a phage that matched Tom's *A. baumannii*, but we couldn't just sit and wait for the answer. We had to move ahead in any ways possible with the administrative reviews and sign-offs so that if phages were found, there'd be no delay starting treatment. We had to be ready to report to the battlefront with new ammunition against this superbug.

We ended our call, each of us with a formidable to-do list. The micro lab had already shipped Tom's isolate to Ry, and the CPT team would now send a sample of his culture to the Navy. I'd start work on the proposal for the UCSD ethics committee. We also needed to get a formal agreement signed to authorize the collaboration and materials transfers between UCSD and Texas A&M and a similar one for the Navy.

Chip's list included following up with the FDA and working up a clinical protocol to establish formal guidelines for administering phages, assuming we got that far. He also needed to get approval from the UCSD biosafety committee. Typically, the administrative process for these various approvals took at least two weeks, sometimes months, but typically they are not a matter of life and death, as they were in Tom's case.

As researchers, we both knew this drill, and we knew the steps and safeguards were there for a reason. The FDA is so often cast as the bad guy standing in the way of innovation, but in fact, those safeguards are important to protect us all. Dying patients and their families are vulnerable. It's too easy to become victims to the likes of snake oil salesmen offering treatments that could kill rather than cure. Worse, in today's world, the profiteers are coopting the language of science and medicine to legitimize their unproven claims and services.

Just as important, treatments that work need to be monitored so their success can push forward randomized clinical trials that will decide if they work on a broader scale. And we need to know when treatments *don't* work so that the deaths of these patients are not in vain and treatment failures aren't repeated. The process took time for good reasons, but it might be time that Tom didn't have. One second too late and it would be game over. And we didn't have any way to know when that second would be.

Two days later, my cell phone trilled. Area code 979. Although it was after eight in Texas on a Friday night, it was Ry. He wanted to introduce me to the four people in the CPT who were trying to save Tom. One by one they introduced themselves: Jason Gill, assistant professor and the CPT faculty member with the most experience doing translational phage research, who was a frequent collaborator of Ry's; Adriana Hernandez-Morales, a doctoral student; and Jacob Lancaster and Lauren Lessor, both lab techs. I thanked them all profusely, and they went to work. So did I. That night I began drafting the proposal to get approval from the UCSD medical ethics committee.

All scientific research done on humans needs to be vetted and approved by an ethics committee. I had worked on scads of these proposals over the span of my career, but this one was different. This time there was only one subject: my husband. And this was a compassionate use request, which meant the team was asking the ethics committee to grant permission to use a therapeutic agent that had not been FDA approved on a patient who was dying.

I wrote about two paragraphs summarizing Tom's case history from the day he fell ill over three months earlier, which included both medevac transports and his downward spiral into a coma after his abdominal drain had slipped a little more than a month ago. I jotted down the timeline and made a list of all the antibiotics he had been on. Tigecycline, meropenem, vancomycin, daptomycin, rifampin, colistin, azithromycin, teicoplanin, metronidazole, imipenem. He was a human pharmacopeia.

Next was the justification for compassionate use. My fingers froze above the keyboard. Paralyzed. I sat there for fifteen minutes, just staring at the computer screen, then the screensaver, which rotated photos from the trips Tom and I had done together. The two of us tracking mountain gorillas in Rwanda last summer. Hiking the Bandiagara cliffs in Mali. Dancing with the Huli wigmen in Papua New Guinea. I couldn't write any more. Not a single word. I felt a paroxysm of fear, anticipatory grief, and panic come over me, and I broke down in huge, shuddering sobs.

Enough.

I sent Chip an email and attached the draft. "Over to you," I wrote.

Spirit lagging, I took a time-out to Skype with Robert, who always left me feeling recharged. He was convinced that the phages were going to work.

"These Pac-Men are going to have a feast!" he exclaimed, rubbing his hands together with anticipation. "And one is like, a super killer, gobbling up more bacteria than any other. In all my fifty years of doing psychic readings, I have never had a more thrilling experience," he told me earnestly.

"Yeah, it's a thrill a minute for me, too," I replied, and we both laughed. Maybe this superbug was going to meet its supermatch after all.

17

A HAIL MARY PASS

February 27–March 9, 2016

Six days into the phage quest, the feverish activity in the phage labs to find a cure for Tom matched that of the TICU to keep him alive. At the center of it all, Tom, still unresponsive in his nest of life-support apparatus.

At rounds the next day, the troupe of TICU docs on Tom's case huddled near the doorway of Bed 11, where Tom lay intubated and wired for monitoring, alive but unresponsive. The group included the attending critical care doctor, Dr. Fernandes; the resident, Eric; the charge nurse, Marilyn; his nurse, Chris; and me. Chris read the summary from the report in front of him, then Eric briefed Dr. Fernandes on Tom's latest cultures and lab values. The team leaned in.

"Creatinine is trending up, from 1.8 to 2.2," Eric said, pointing to his laptop. His tone communicated escalating concern. The normal range is 0.6 to 1.2 in adult males. Dr. Fernandes leaned over to view the past week's trend in creatinine levels, a marker of kidney function. The bar graph's steep upward slope was not good.

"Better call in nephrology," he told Eric. I knew what it meant. Tom's kidneys were starting to fail. And given his condition, dialysis wouldn't buy him much time, if any. Since Tom was already on life support for heart and lungs, dialysis would signal a terminal trifecta of system-wide organ failure—the beginning of the end.

Ordinarily, on rounds the team would stay just long enough for the update and then move on to conduct rounds at Bed 12. I was a fixture there by now, often asking questions to clarify technical aspects of Tom's condition, his medical treatment, and care. But today, Dr. Fernandes and Eric looked first at each other, then to Marilyn and finally at me, with a question of their own.

"There's a rumor you and Chip are planning some sort of experimental treatment involving viruses," Dr. Fernandes began, his tone cautious but curious. "Can you fill us in? None of us have any experience with this, and we need to be in the loop." All eyes turned to me.

For a moment, I was caught off guard—embarrassed that they might think I was questioning their abilities. But I saw in their eyes that they were desperate to save Tom and frustrated that they'd run out of options. There wasn't much about modern medicine that the doctors in this world-class hospital didn't know or weren't experienced with, but the blind spot was understandable, since phage therapy had been sidelined decades ago in Western medicine. That might be about to change.

I explained that Chip and I had only enlisted the help of CPT and the Navy less than a week ago, and we didn't yet know if they'd find any matching phages, or how long it would take. Dr. Fernandes and Eric listened, their expressions reflecting keen interest and natural skepticism. As a scientist myself, I knew that feeling.

"I know it's a long shot," I said. "And I know it's risky. But I don't see any alternatives being brought to the table."

Dr. Fernandes nodded slowly. "Neither do I. To be honest, we are running out of options to turn your husband's case around."

It would be a Hail Mary pass with the quarterback blindfolded and less than a minute left in the fourth quarter, he added, but it was worth a try. "If you can pull this off, it may benefit a lot of other patients. So, you and Chip have my full support."

I thanked him and promised to update his chief, Dr. Atul Malhotra, as well as the director of the TICU, Dr. Kim Kerr.

Left to ourselves, I brushed Tom's hair gently and rubbed his forearms and legs with an extra thick cream. Then I took a pumice stone and exfoliated the bottoms of his feet, rubbing off layer after layer of calluses and peeling skin. "You're finally getting that pedicure I've wanted you to get for years," I told him, pretending he could hear me. Maybe he could.

At noon, I prepared to leave Tom in Chris's capable hands, knowing that in a few minutes, two of our students would arrive to sit with him. Carly's husband, Danny, had sent out the call online for this bedside vigil and used an online calendar tool for scheduling. The response from all corners had been overwhelming, and friends, family, students, and others showed up without fail for two-hour shifts, throughout the day and evening.

It was impossible to know whether Tom, in a coma, was in any way aware of their caring presence or anything I might do or say. But a few studies had suggested that comatose patients could sometimes hear, and that the sound of their loved ones' voices aided their recovery. And Robert insisted that their presence and interactions were a vital connection to Tom, a grounding, human tether through his otherwise utter isolation in the mass of tubes, cables, and high-tech medical paraphernalia that enveloped him. Martin, the holistic healer, agreed. However difficult it might be to reconcile with evidence-based metrics, Tom had rallied unexpectedly in Frankfurt when his daughters flew over to be with him, even when most clinical signs indicated that he was dying. There had been other moments, too. Nothing you could quantify, but honestly, if modern medicine was down to digging through sewage for experimental possibilities, if some bedside company and one-way conversation held any potential for good, I didn't need FDA approval for that.

Tom stirred briefly and moaned softly as I kissed him goodbye until tomorrow.

On my way out, Rosie, the TICU housekeeper, entered Tom's room with her cleaning cart, as she did nearly every day. I had recognized her hobbled walk heading our way from way down the hall. I greeted her with a weak smile.

"It's all yours, Rosie," I told her, watching mindlessly as she swept the floor of Bed 11 clean, emptying into a biohazard bin the dustpan full of syringe caps, crumpled Kleenex, and the layers of dead skin that continually shed from Tom's feet.

Rosie's eyes met mine. "I'm praying for you and your husband. You are both so full of life," she said gently. Her kindness, and just the mere thought of Tom as "full of life" made me choke back a sob.

The next day, Tom's weakening state due to his lack of nutrition became a new focus of urgent attention at rounds. The tube that had been inserted through his nose when he arrived at the TICU wasn't up to the task of handling his nutritional needs now. He was, in any practical sense, a starving man. However, the bigger, better type would mean another tube inserted in his abdomen, yet another potential site for infection that could instantly trigger more sepsis. After agonizing over what to do, I signed the consent authorizing interventional radiology to insert a new G-J-tube in his jejunum the next day, after no amount of hand-squeezing could coax a response from Tom.

Two days after the procedure, I sat with Tom bedside. He lay quiet, still unresponsive, but when I arrived I was dismayed to see that he was in wrist restraints. Not even fully conscious, he'd put up a fight against the vent tubing, and this was the only way to keep him safe. I accessed Pandora on the laptop and selected some of his favorite meditation music, Native American drumming. The warm tones of the drumbeat filled the room, a welcome contrast to the cold, syncopated beep of the heart monitor. Then suddenly, the blare of the cardiac monitor shattered the peace. Tom's heart rate shot past 130, and his oxygen level dropped below 90. I smashed the call button. Chris had just come on shift and came running down the hall. Tom's breathing had become rapid and shallow. His face turned red and shone with fresh perspiration. Chris and I looked at each other, read the signs, and came to the same conclusion at the same instant: more septic shock.

All along, there had been trouble adequately draining the infection sites in Tom's gut. That afternoon, new blood tests, lab cultures, and CTs were conducted. A few days later, it would show that a fungus—*Candida glabrata*—which had been detected in the pseudocyst from the beginning was now present in Tom's blood. Somehow the fungal infection had breached the walls of the pseudocyst and was spreading. This meant that Tom had candidemia, which Davey reminded me had a 50 percent mortality rate. Tom was now under siege by bacteria and a fungus, while we were trying to save his life with billions of viruses. For an infectious disease epidemiologist, life seemed more and more like a cruel joke.

••••••••

Some mornings, the hospital's sunny atrium and palm-lined promenade was invisible to me in my rush. It could seem overly earnest, too sunny, too optimistic, when I was hurrying through for the umpteenth time up to Tom, who was listing more to the dark side. That doorman with the snazzy double-breasted suit who used to look so welcoming was starting to make me think of a pallbearer. I'd walk in as other families walked out with their loved ones newly discharged, holding balloons, and I tried to be happy for them, but it was hard wondering if Tom would ever get out. For all the attention paid to time, all the ways the TICU staff sliced and diced and measured it by heart rates and shift changes and scheduling and bowel movements, in a real sense, time stood still. It had been three months since Tom fell ill in Egypt, two months since he'd been airlifted from Frankfurt to Thornton, and it felt like forever. Now, suddenly, the promise of the phage cocktails had quickened the sense that something was finally happening. But until the phage arrived and Tom could be treated, nothing *could* happen. Instead of feeling time oozing formlessly around us, it was now electric, supercharged, tension mounting in the wait for someone to flip the switch. It had been two weeks since we'd hatched the phage therapy plan. Depending on my mood, the days felt like either a countdown to blastoff or the last dark minutes of the Doomsday Clock.

When I arrived at the TICU the next day, Tom lay in Bed 11, so still that I had to check if he was still breathing. Yep. The cardiac monitor showed his heart rate at 113, tachy, but stable. Blood pressure 90/65. Not great, but no change since my usual five a.m. call to the nurses' station a few hours earlier.

"Has he been awake at all since shift change?" I asked Chris, who'd been assigned to Tom for a second day. Chris had a warmth and presence that never failed to transform our sterile stage into a human one. So much about the ICU focuses naturally on the essential—"keep them alive"—but Chris always went a step further, making minute adjustments in care to not only keep Tom alive but also keep him comfortable, actively make him better, even in small ways. He always took the time to explain ICU lingo to me in a way that didn't feel patronizing. Chris had just finished brushing Tom's teeth with a practiced hand and was getting ready to turn him on his side, as he was required to do every two hours to ensure that he did not get any "pressure wounds." Bedsores.

"No change since you were here yesterday morning, I'm told," Chris said solemnly. Still unresponsive.

When he'd been admitted to the TICU in December, Tom had confessed that he hated to go to sleep at night because he was afraid he'd go to sleep and never wake up. It seemed that most people who died in the ICU died at night—literally during the graveyard shift. He tried not to pay attention to the code blues and the gurneys that rolled in with skeletal people lying on them and rolled out with covered-up cadavers, and the smell of frankincense in the air, from the priest who came to give some their last rites. But it was hard not to watch. "We're the nearly deads and the newly deads," he'd declared. The fine line between the two was becoming more frightening by the day.

I plunked down in the chair beside Tom's bed, exhausted. "No change" left us in limbo, but suddenly the sight of Chris tending calmly to Tom, and Tom resting quietly and not dead, let me step out of panic mode. That afternoon, there was no way to sugarcoat the results from the latest CT. A new fluid collection was expanding in the back of Tom's pancreas, and another one where the new feeding tube had leaked. Two more drains were inserted, for a total of five. Tom now looked like a pincushion. He had lost 100 pounds, and his skin took on the pale and waxy look of a corpse. The hum of the monitors that had been so annoying before had become the only note of reassurance that he was, at least, still alive. All hope now rested on the phages.

18

PANNING FOR GOLD

March 1–March 11, 2016

Wherever you find bacteria, you find the phages that prey on them. If you want to find phages with an appetite for intestinal bacteria, a pile of poop is a good place to start. For *A. baumannii*, you go low. Or, as one Navy scientist summed it up, "some of the worst places you would ever want to go." These include sewage treatment plants, standing cesspools, or dumps filled with dirty diapers and other fecal-tainted debris, rotting garbage, the occasional dead animal, and maybe wastewater runoff from a local hospital or animal farm. The Navy's environmental phage sources include ships that travel internationally. Since some nasty superbugs have been found in the pipes and sewers connected to the NIH's hospital in Bethesda, they also collect samples from nearby sewage treatment facilities.

The other place you can find phages is the opposite of these retch-worthy locales, where the dirty work has already been done. These were the spotlessly clean, well-ordered phage "libraries," a walk-in fridge in some low-traffic corner of a microbiology lab. In this library, the collection consists not of volumes but vials, most no bigger than your pinkie, labeled and stored for easy reference and access—just what we needed now.

Whether sifting through library or lagoon, the search for the particular *A. baumannii* phages active against Tom's specific isolate was going to be painstaking. With time such a critical factor, we primarily focused on sources in the continental United States, where transport would be easiest. And we got lucky. Between Texas A&M's CPT and the Navy's Biological Defense Research Directorate (BDRD) lab, we'd tapped into some of the most experienced and respected phage scientists on the planet.

Ry had devoted his long career to studying how phages manage to detonate bacterial cells (lysis) when replication is complete and the new progeny are ready for release. Tough bacterial cell walls are able to withstand enormous pressure, but somehow phages are able not only to cause the infected bacterium to explode but to do it on a timer, chosen to optimize the number of virions released.

At the Navy's lab and NIH before that, Dr. Biswajit Biswas, Carl Merril's former protégé and now Theron's phage team leader at the Navy lab, had worked for decades to develop ways to effectively select the best phages for treatment. Among other things, he had helped develop a faster system to grow cultures and phages, using an automated incubator with a sophisticated camera and computer to monitor and analyze the data in real time. All this meant the lab could judge the therapeutic efficacy of different phages and different combinations of phages to kill target bacteria in hours instead of days. In recent years, the Navy's extensive work on therapeutic uses of phages to treat wounds had involved *A. baumannii*, with the intention of eventually using phage therapy to treat multi-drug-resistant infections in service members or VA patients. But so far, they'd only tried it on lab animals. No humans. Tom would be the first.

Both teams were packed with skilled, passionate scientists, each with a distinct focus, history, and strengths that complemented the others. Suddenly, it felt like a dream team had materialized out of nowhere.

Their mission was straightforward but not simple. They had to find the phages that would target Tom's multi-drug-resistant strain of *A. baumannii*, then grow the phages in quantity, purify them, and get them to the UCSD's investigational pharmacy to be prepped. Ordinarily, all of this would take weeks, but with daily, sometimes hourly, updates and consults with Chip and me, the team knew Tom was failing fast. They had to find ways to hurry the process without sacrificing quality—and Tom. They couldn't hurry Mother Nature; phages need time to replicate, about twenty to forty minutes once they infect a bacterial cell. Harvesting them, repeating the propagation process, and the rounds to purify them eat up the clock. So everyone opted to work harder and faster to close the routine gaps, doubling up on the numbers of assays underway at one time and working through nights to eliminate downtime between steps. There'd be time to sleep later, Theron joked. For now, the race was on, from bog to bedside.

One team in Maryland, the other in Texas, the medical team and the rest of us in San Diego, and others online were different time zones and miles apart but up against *A. baumannii* in a single race against the same clock. If Tom survived the first round of phage therapy, once the battle was underway, win or lose, we were all on the same team.

While Chip and I forged ahead on the regulatory and administrative fronts, the CPT and Navy teams set out to find and prepare the phages as rapidly as possible. Both labs had a combined collection of hundreds of phages that were active against a number of pathogens, but at the time I contacted Ry, they had only a few *A. baumannii* phages on hand, most left over from various phage hunts in the past few years.

"We're lucky that we never clean out our fridge," Jason quipped. Pickings were slim simply because the Texas group hadn't done a concerted phage hunt for *A. baumannii*—until now.

The Navy, on the other hand, had been doing precisely that. Many sailors and other service members had returned from Middle East wars with Iraqibacter infections. They usually weren't as virulent as Tom's infection, but their existence had spurred the Navy's search effort, and we were now the beneficiaries. The Navy had extensive libraries of a few thousand phage specimens from environmental sources—those collected from the muck. Of the some three hundred partially characterized phages in its collection, one hundred fifty were active against various strains of

A. baumannii. Since *A. baumannii* phages were more finicky than most, there was no telling if any would match Tom's strain.

Biswajit was known among peers as the Phage Whisperer, or as Carl Merril put it, a kind of phage matchmaker who had an intuitive sense of how to wrangle phages to get the best results. From a family of veterinarians in India, he had begun a medical career as a vet himself there. But he was also a born tinkerer and in his long career as a scientist had earned a reputation as an inventor. He had a sixth sense when it came to choosing phages that kept the selective pressure on an evolving bacterial infection.

In a way, Biswajit had been preparing for this moment for nearly twenty years. He had worked with one of the pioneering phage companies in the US in 1994, and although his animal studies were promising, like Carl, he hadn't been able to get any traction for phage therapy in the medical community at the time. Although he had remained convinced that phage therapy could be a life-saving medicine, eventually he shelved the work. With the Navy BDRD, his work with phages had included those that kill anthrax, a bioterrorism concern. When the call came to develop a phage cocktail for use in Tom's case, the long hiatus was over. His time had finally come.

The Texas and Navy labs differed in their setups and processes to get the job done, but once you had phage in hand, the basic steps of a phage selection process began the same. You screen the phages to be sure that the phage you carry forward is a lytic active killer phage and not a temperate phage that embeds its genetic material into the bacteria and hits the snooze button. The thinking was that temperate phages could make matters worse, because they often transmit toxin genes as well as antibiotic resistance genes they've picked up from other bacteria.

Once you've isolated a phage that's active against the target bacteria, you have to grow lots more of it to obtain enough for a therapeutic supply. Since phages can only grow by infecting and killing bacteria, you must mix the phage and bacteria together in a culture. Nothing about the basic interaction between phages and bacteria had changed since Félix's day, when researchers worked with a clutter of Petri dishes to culture organisms. In contemporary labs, though, where work involves sophisticated technology and is on a larger scale, Petri dishes are now microtiter plates the size of an iPhone, each with ninety-six dimpled wells for culturing bacteria and isolating phages, then culturing again to amplify the phages' effective against the target bacteria. Walk past a few of the clear-covered plating panels laid on a lab counter, and it's a little like an aerial view of Midwestern farm fields in late spring, a blanket of neat, dotted rows of carefully cultivated crops, row after row after row.

When the phages finish growing, you can see the glossy plaques where they've destroyed the bacteria. Researchers then harvest the phages with a glass pipette. After growing and harvesting many rounds from the bacterial "lawn" on the dimpled plating panels, you've got a flask full of phages—as many as ten billion per milliliter. But the other product of phage growth is a lot of dead bacteria, cell parts, and environmental flotsam and jetsam. Part of that cell debris is endotoxin, a toxic part of the bacterial cell wall that must be separated from the phages to minimize the risk that it might trigger septic shock in the patient. Purification also removes other potentially harmful residuals from the process. Carl later told me that in several of the early phage therapy experiments in the 1930s, phage concoctions may have killed more people than they cured, because no one knew that they needed to be purified or how to do it. These days, purifying the phages was the hardest part.

Typically, researchers put the crude phage preparations in a large centrifuge tube and then use a powerful spin cycle that forces the phage particles and other debris to sink to the bottom of the tube, leaving the debris behind. Ry fondly described his vintage 1970s centrifuge as something like a souped-up Cadillac engine in an old Maytag washer. When it was cranked up to full speed, you didn't want to open the lid on that g-force. Once spun, the phage concentrates are dissolved in a sterile liquid, which must be done carefully to avoid damaging the phages themselves. The next step is to mix them with an organic alcohol to remove the endotoxin. The final phage prep would be tested again and treated again, if needed, to bring the endotoxin levels down.

There are a few different options for purifying phages, and the Texas and Navy labs each had their own processes, but ultimately, they'd need to meet the FDA requirement for safe levels of endotoxin in a product for human use. No one really knew how pure the phage preps needed to be for treating humans safely, so the cleaner the better. We wanted squeaky clean, virulent phages, ready to take *A. baumannii* down fast, and preferably an assortment that had evolved to overcome *A. baumannii* defenses by targeting different receptors. The more phages found that were active against Tom's *A. baumannii* and the more varied their weaponry against it, the better. That would keep the bacteria engaged in a constant battle on different fronts, meaning its own pathogenic arsenal would be spread thin, increasing the chances that the phages could exploit a vulnerable flank.

In the days ahead, as the labs began their work of hunting, harvesting, and processing phages, Chip and I each had a list of administrative checkpoints requiring procedural paperwork, permissions, clinical plans, committee reviews and approvals, and legal papers that ensured that those who supplied the phages would not be held legally culpable if Tom died. We used every connection and every ounce of goodwill that we had at the university to get the approvals we needed. The paperwork for this typically took weeks or months, but we managed obtain it within two days, thanks to countless hidden heroes, those who do the chop-wood-carry-water administrative work from the inside.

In the middle of the night in early March, three weeks into the phage hunt, Jason emailed Chip and me with both good news and bad news from Texas A&M. They had tested a small sample of the Belgian phage, but unfortunately, the phages weren't active against Tom's *A. baumannii.*

The good news was the team had found four phages that were active against Tom's isolate. One was a phage from AmpliPhi Biosciences, a phage therapy startup—in San Diego, no less—specializing in the development of phage-based therapies for clinical use. AmpliPhi hadn't surfaced in my search because I'd focused on labs that were working on *Acinetobacter* phages. AmpliPhi's focus was not on these phages, but they happened to have one that originated from a patient in Australia, which they gave to CPT. When Ry contacted the AmpliPhi CEO for permission to use it in Tom's cocktail, he instantly gave his okay and, understanding the urgency, required none of the usual paperwork.

Three other phages were found *de novo*, from environmental samples pooled from sewage, soil, watering holes, and poop collected from swine and cattle barns near their lab at College Station, Texas. In total, these three new phages and the phage from AmpliPhi were the first phages selected for the Texas cocktail. Even through email, Ry's reply in the morning bristled with excitement at the realization that his lab had identified three new *A. baumannii* phages in just a matter of days. All they needed now was to purify and brew them. Texas tea.

••••••••

On March 11, not quite three weeks from the start of the phage hunt, we received an email from Biswajit at the Navy lab. Pending the formal go-ahead, Biswajit and his lab staff had conducted preliminary tests of the *A. baumannii* phages in their collection against Tom's bacterial isolate, now formally named TP1. They quickly identified ten highly virulent phages that killed Tom's bacterial isolate. Within eighteen hours Biswajit had carefully selected the most virulent four phages they would carry forward to make their first cocktail.

Theron had some red tape to wrangle, but he found an ally in the BDRD director, Dr. Alfred Mateczun. Mateczun was a retired Navy captain who was also an MD and had a reputation for integrity and decisive action. He had been heavily involved in responding to the 2001 anthrax attack aimed at several US politicians, shifting from research to sample analysis for almost one year. When Theron contacted him about Tom's case, Mateczun's response was characteristically blunt.

"Well," Mateczun said matter-of-factly, "he's been in a coma for what, six weeks? He's going into organ failure, and may die soon." He boiled the rest down to two questions.

"Do we have the bacteria from the infection?"

"Yes, sir," said Theron.

"And do we have phage that will kill it?"

"Yes, we do, sir."

"Okay, we have nothing to lose. Send the phage."

With that, Mateczun hit the Go button. The Navy had answered our SOS.

Now, as the Biolog automated system incubated, monitored, photographed, and charted the phage-bacteria interaction every fifteen minutes, the Navy lab harvested and cultivated the fittest of the fittest phages. From small-scale harvesting of the phages from plaques, the lab infected flasks full of *A. baumannii* to produce 3.6 liters—almost a gallon—of the raw mixture. A sixteen-hour spin cycle in the centrifuge produced about 10 milliliters—two concentrated teaspoons of the active phage preparation, called lysate—and they were good to go.

I'd only worked with phages in one undergrad class, and it wasn't particularly memorable—except for the intense smell of those bacterial cultures we'd used in phage experiments. That was unforgettable. The Petri dishes were incubated at body temperature in what resembled a microwave. When you opened the door, a waft of warm air rushed out and you were overcome with an odor something like a mix of old socks, armpits, and decay. When I heard one day that Biswajit had joked about how smelly the lab had become, I had to laugh. I remembered the long nights babysitting putrid plates of bacterial broth. But to a diehard phage researcher, that's the smell of potential. And Biswajit was a guy who lived and breathed phages. A sewage treatment plant in his community wasn't far away, and he remarked that on some days, when the blistering summer temperatures were just right, he'd catch a whiff and think, "Ah, it's a perfect day to hunt for phages!"

••••••••

As I worked through the formal paperwork required of me as Tom's wife, small things constantly reminded me of my strange double life as scientist and spouse. I was now over any hesitation about pressing in my professional capacity for information or procedural responses to expedite the administrative approval processes. Meanwhile, every form approving treatment or interventions for Tom called for me to sign as his spouse, "the patient's wife." Just seeing the words made my stomach clench. Every minute of every day was a walk on a high wire, emotionally. I could put one foot in front of the other just well enough—as long as I didn't look down. Focusing strictly on what needed to be done next seemed to work: Scientist Me could maintain a semblance of objectivity and detachment, and Wife Me was able to stop short of the unthinkable outcome of Tom dying.

I remembered the dream, the nightmare really, that I'd had only a few weeks ago. I'd waded in that putrid, swirling swamp, hunting for phages like an old California prospector panning for gold. I'd woken in a sweat, seeing the cracked bedpan from Luxor as a bad omen. Now the dream was rewriting itself, the image transforming before my eyes in real life, and in ways I could never have imagined. I wasn't alone. These wonderful people, this volunteer army of strangers, had stepped forward from far corners of a world that had been invisible to me before. They were the gold. And they were working day and night to find the phages that would prove golden, too.

I knew how labs operated, the demands on staff and resources, the precision of the tasks, and extensive planning that goes into their scheduling and the work itself. This had to have been a nightmare for them, too. A project dropped in out of nowhere, on a desperation deadline, disrupting everything and demanding even more. All that, and with a human life at stake. I could relate, but then, this was my husband. Tom was a stranger to them. And this was a huge gamble, with all odds against success. Professionally and personally, it had all the signs of a heartbreaker. And yet they stepped up.

From time to time, someone would share a glimpse of how personal the project had become to them. Each of them had the dedication it takes to work for years on something that you hope will eventually save lives. But it's not often in research labs that the opportunity presents itself for you to be a first responder in an effort to save someone's life *right now.*

After one long stretch of particularly frustrating work to tease out differences between the phages she'd found, Adriana later said, she suddenly hit the jackpot and found what they needed to add the fourth phage to the arsenal against Tom's *A. baumannii.* She was ecstatic.

"I named that one *Mago*," she said. "*Magician* in Spanish, my native language."

Perfect. We needed all the magic we could get.

The same day, Jacob and Adriana in Texas texted me a photo of the package as they waited for the special pickup for a same-day delivery to the UCSD lab. I knew they had just pulled several all-nighters. With phages en route from Maryland and Texas, it was our own California Gold Rush, westward ho.

As the teams worked as fast as they could, the vigil with Tom continued at the hospital. With the photo from Jacob and Adriana fresh in mind, I put a damp cloth on Tom's forehead to give him relief from yet another low-grade fever. He had not been awake for days.

"Hold on, honey," I said, leaning close, "the phages are coming!" The vent and pressor settings were now set to high. I squeezed his hand as hard as I dared. He did not squeeze back.

19

JOURNEYING

March 12–15, 2016

I think his spirit is traveling," said Frances pensively, as she poured herself a cup of tea, a brew she had concocted from black Chinese herbs with unpronounceable names. A pungent smell like burnt licorice filled the kitchen and wafted after her as she carried two steaming mugs toward the couch. She held one out to me. "Here, Steff, have some tea. It's good for you."

"Thanks, Fran." I smiled at her from my favorite spot on the sectional sofa. "I hope it tastes better than it smells," I said with a wink, and crinkled my nose. Frances had taken a course in herbalism and was planning to enroll in a traditional medicine program. She knew her herbs. I took a small sip, which tasted like treacle. Not bad.

"I think he is, too," said Carly, who had emerged from the bedroom to plop down beside her sister and me. She released her hair from a long ponytail and gave it a shake.

Carly and Frances had returned a few days ago and we took turns at Tom's bedside, waiting with bated breath for the phages to arrive. At rounds earlier that morning, the mood had turned from anticipatory to bleak. After reviewing Tom's chart, Dr. Fernandes looked me in the eye and asked if the three of us could attend a family meeting later that afternoon. We returned home for a few hours, leaving Tom with our friends Chuck and Judy.

We were all coping with Tom's illness in different ways.

Every morning, Carly would retreat into the guest bedroom, insert earbuds, and listen to a recording of shamanic drums as she kept one question at the center of her mind: *How can I help my dad?* She described her journeys as traveling to a world bathed in light where she met with one of a council of guides. In her most recent journeys, she said, she would immerse herself in something like a giant centrifuge that spun off her sadness so that she would be ready to help her father without the burden of her own feelings. The drone of the drums would then lead her to a place where her dad was lying in bed. She would help him up, and together they would walk to a bench overlooking the Pacific Ocean, where they would sit and talk.

"Maybe Pops is wrapping up, saying goodbye," Frances suggested now, refilling my cup. She had just finished meditating, too, and she spoke in a calm voice that hid the fact that she was more upset than she was letting on. I had spent the last few hours pacing the floor, double-checking the online tracking number for the phage delivery to the hospital pharmacy and conferring with Chip. I could not sit still unless I had a kitty on my lap. Thankfully, Paradita had complied and was purring and kneading my velveteen bathrobe; her claws were like tiny scalpels.

Carly and Frances were seated on either side of me on the sectional. We gazed out at the mesmerizing view of the ocean through a floor-to-ceiling window that spanned the back wall of our house. Tom and I had spent hours watching the sun set over the ocean from our backyard. He was always searching for the green flash, a rare but natural phenomenon at sundown that is caused by refracted sunlight. I had almost finished my tea and tilted my mug so I could pluck out a piece of broken leaf that had clung to the bottom.

Within the hour, Chip called me with bad news. Before administering the phage, we would need to reduce the endotoxin concentrations in the phage preparations further, to as low a level as possible. The FDA had guidelines about this for other kinds of pharmaceuticals and vaccines but not for phage preparations.

"They have quite a bit of experience in this area, and I've been talking with them about what concentrations they think we should shoot for," he said.

The phages that were to be used to treat Tom's infection had been grown in large-scale cultures of his own *A. baumannii* at both Texas A&M and at the Navy. When these bacteria were destroyed as the population of phages was expanded in the lab, they released one of their key components that the human immune system recognizes when it is invaded by bacteria: endotoxin. Endotoxins are important triggers of the immune response and one of the major drivers of fever and low blood pressure in severe infection. The FDA was concerned that if there were significant residual levels of endotoxin in the phage preparations with which we planned to treat Tom, it could lead to septic shock. I knew from my crash course in phage biology over the last few weeks that the majority of deaths during phage therapy had been attributed to endotoxins that had not been adequately removed in the purification process. However, the phages themselves would also release endotoxin originating from the bacteria they killed within the body. That could cause septic shock, too, so Chip and I both wondered if concerns about the endotoxin that might still remain in the phage preparations warranted a delay in the treatment. In the end, Chip decided we'd better bring the endotoxin levels down further just in case. Better safe than sorry. Or dead.

Texas A&M had already shipped their phage preparations to the UCSD pharmacy, so their ability to conduct purification assays was limited to the individual phages, and not the cocktail. What was worse, their lab and the Navy had used a different technique for the phage purification, and since administering phage systemically in humans was a new approach, it was not clear which—if either—of these approaches would be workable.

What now? These new hurdles would delay the start of treatment by some unknown margin.

No sooner had I hung up the phone, despondent, when Ry called me.

"I heard about Chip's discussions with the FDA about endotoxin levels in the phage preparations," he said. "This is exactly the kind of problem others have run into when attempting phage therapy. I am not sure we can overcome it in time to save Tom, but we have to try. As luck would have it, I know a team of phage researchers in your own backyard, at San Diego State University."

Again, I was stunned by the serendipity that led us to individuals with a particular expertise, some of whom, for a variety of reasons, had simply not shown up in our earlier searches. It seemed more than lucky that some of these people were right here in San Diego, engrossed in work that had nothing to do with using phages to treat sick people. Until now.

"Talk about stars in alignment—what are the odds?" Ry said, genuinely amazed.

The miracle of the moment was to be undertaken by Dr. Forest Rohwer and his research team at SDSU—his lab and life partner, Dr. Anca Segall; their postdoctoral fellow, Dr. Jeremy Barr; and Sean Benler, a doctoral student. Forest's field was the eco-biology of phages, a field he and Jeremy had put on the map. Generally, their research was focused on phages as a model for studying DNA, genetic engineering, CRISPR, and other basic science—not phages for therapeutic use. Forest was known for his success in "scrubbing" phages clean even when the details made it dicey. Anca, a molecular biologist who was also a SDSU professor, headed up her own lab, and the two collaborated frequently on studies of marine phages. As it happened, Jeremy was an expert in the highly specific endotoxin removal method that was needed, and he had just recently concluded a year-long project using the process, so the lab was set up for this specialized work.

"I've left them a message to see if they would step in to run an endotoxin purification assay on our phage cocktail," Ry said. "They're good people. They'll need to have a higher-quality endotoxin kit than the one we have here. Cross your fingers that they can help; otherwise you'll need to find another lab that can do the assay on the fly, or ship the phage preparations back to us so we can re-scrub them."

Any delay pressed Tom's luck, and the repurification step only added to our fear that Tom wouldn't be able to hold on long enough.

"Ry, thanks so much for stepping up to the plate yet again," I told him, exhaling loudly. "I'll call your colleagues at SDSU, too. I hope they'll agree to help, because time is running out." I looked at my watch. Eleven a.m. Literally the eleventh hour. "We're headed to the hospital again this afternoon for a meeting with the doctors, and I have a feeling I am not going to like what they have to say."

I was right about that. I left a voice mail message for Forest. After lunch, Carly, Frances, and I set out to the hospital, where three familiar faces from the medical team met us in a small conference room.

"The purpose of the meeting is to discuss next steps," said Dr. Mims, who chaired the meeting. He ran a hand over his head, which was balding prematurely. "I've cared for Dr. Patterson off and on for the last several weeks, and he isn't getting any better. In fact, I think everyone here would agree that his trajectory has worsened."

We know this, I thought. *Get to the point.* He did.

"No one can say what will happen, of course," Dr. Mims went on. "But in my experience, a patient with this kind of clinical profile is not going to improve. We need to know if you want to start planning for kidney dialysis or not. If you start dialysis and he recovers, he may need it forever. And even if he recovers, which I don't think he will, he will need over a year of intense rehabilitation."

"But he's not in kidney failure yet, is he?" I prodded. "I thought you didn't start dialysis until his creatinine was much higher." A sick feeling crept into my stomach, like the treacle tea had curdled.

"Nephrology rounded on him earlier, and they don't like the way things are headed," Dr. Fernandes added gently. "His creatinine is now over 3.5. He will need dialysis very soon."

I sat back to process this, closing my eyes. Lungs, heart, and now kidneys, all failing. This was my worst nightmare; Tom was on the verge of multi-system organ failure, the phages weren't ready, and we were basically being asked whether we wanted to pull the plug.

"If he needs dialysis, and we decide not to give it to him, will he die?" Frances asked, pausing to swallow hard. Dr. Mims looked at her and nodded. "Yes. And soon, I think."

"So," Carly said, "would it be possible to just take him home to pass so that he doesn't have to die in the hospital?"

Dr. Fernandes looked at her calmly and replied in a soft, practiced voice. "He would only live for a few minutes if we took him off the ventilator and stopped the pressors, since they are what is keeping his heart pumping," he replied.

Carly's chest deflated. "So, there is no way he can talk to us, just one more time?"

"I'm afraid not," said the third doctor. His face was a blur to me, and I realized it was because I was crying.

Time to step up, I told myself. "We have not given up yet," I said firmly, with a cool confidence I did not possess. "The phages are almost ready, and while we do not yet have FDA approval to proceed, we expect it will be granted very soon. So, if his kidneys need dialysis, please give it to him."

The girls nodded their assent.

"Understood," said Dr. Mims. "We will leave the phage plans to you and Chip. But I must warn you that there is precious little time left to turn your husband's course around." And with that, all three doctors stood to leave, their white coats flapping and reflecting the fluorescent light, which suddenly seemed unduly harsh.

Outside the conference room, I checked my cell phone, which had been set to vibrate. There was a voice mail message from a local caller. It was Forest calling from SDSU. He had received the messages from Ry and me and had cleared the decks at his lab to conduct the endotoxin purification assay. I was so overcome with relief that my knees almost buckled and I held on to a hospital rail to catch myself. An orderly in blue scrubs who was passing by paused to see if I was okay. I waved him on with a thin smile. There was still hope.

A few phone calls later, I had briefed Chip on the good news and located the Texas phage cocktail, which had just been delivered to the UCSD investigational pharmacy unit. Not trusting a courier to get there quickly enough, I asked my colleague Natasha to pick up the box and drive it over to Forest's lab, so they could start prepping a sample for the assay.

"Remember, the phages are packed on ice and need to be kept at four degrees Celsius," I instructed her, "or they will die." I almost corrected myself as I said this, because technically viruses aren't living or dead. They exist in a form of limbo until they come into contact with a bacterial host cell. Like a coma.

A mix of factors added to the complexity of Tom's situation as Chip and the team worked to treat the underlying pancreatitis and the riptide of complications that yanked him from one crisis to another. Practical challenges loomed large. For one, no one here was trained to do phage therapy. There wasn't an established protocol or even a rough draft—yet. But Chip was on it, brainstorming for hours on the phone and by email with Carl Merril in Bethesda and Maia Merabishvili in Brussels, who, in turn, consulted others on details from their varied experiences involving dosages, frequency, and other tips on how best to administer the phages. But as Jason, at the CPT, later put it, the specs were "all over the place."

While the SDSU and Navy labs worked to purify their phage cocktails to the FDA's specifications, Chip carefully considered all the data, weighing the evidence and the options against the risks and possible outcomes. "Fully colonized," Tom's infection had spread throughout his body, but there was no way to tell precisely every place it festered inside him. Should the phages be injected into the catheters in his abdomen, which would be closest to the source of the infection? Or should they be administered intravenously? How should the pharmacy prepare the phages? What concentration should be used, and how much and how often should a dose be given?

Flooding Tom's system with the phages intravenously meant that, hopefully, the phages could be carried throughout to reach any hidden bacterial reservoirs. Phages are inert, so there's no great mobilization that they initiate as they come aboard. But their infinitesimal size allows them to diffuse easily through the human body. However, an IV infusion was also the riskiest move in terms of endotoxin exposure as the phages burst through the bacteria and left the debris free-floating throughout his system, which could drive Tom's immune system into yet another episode of septic shock.

Getting the dose right is critical to minimize risk to the patient while still maximizing the element of surprise for the bacteria. Shock and awe. Then repeat. *But how do you know the dose to administer when the drug is alive?* Phages aren't drugs; they're viruses, organisms on a seek-and-destroy mission of their own, and they—like their prey—respond in real time to modify their equipment for attacking their preferred host bacteria.

"Phages are the only medicine that multiply, the only medicine that mutates as it works," Ry said in a light moment, with a touch of reverence. I thought of all the years that scientists like me had spent in pitched battle with deadly viruses precisely because they're so good at being such killer adversaries. Now I could only pray that these phages could use those skills to outwit *A. baumannii.*

"The experience with human phage therapy in the US is a mixed bag," Chip explained at rounds that weekend in the tone he saved for serious understatement. Animal experiments were one thing. And different kinds of phage applications involving patients—topical treatments, for instance—were another.

If he survived the initial treatment, what next? Which phages would be needed for continued treatment, in what quantity, and for how long? How should the cocktails be modified to match the inevitable resistance that *A. baumannii* would develop, when there was no way to predict the bacteria's next move? The concentration, volume, and frequency of the doses would affect how long the initial supply would last. The unknowns clearly outnumbered the knowns. And yet, Chip plugged away at the puzzle with his characteristic capacity for complexity and calculated risk. He'd earned his reputation as a maverick to be trusted, a brilliant mind with a strong moral compass. He was willing to take risks that others might not, but always with a patient's best outcome as the end that justified the means.

Reading the risk, mining the data, and getting it right. At the moment, Tom's life hung in that balance.

****

It wasn't like Forest to call late at night, especially on a Sunday, but when he called Jeremy Barr, his postdoc fellow, and told him about Tom's situation and the urgent need for their help to purify the Texas phages, Jeremy understood the rush. It was all hands on deck. Forest had called with a number of questions, but they boiled down to this. Was Jeremy confident that their particular phage prep process could purify the phages to the FDA level for *human* use? That hadn't been done before. And could they do it in twenty-four hours?

Jeremy did a quick mental calculation of the hours required for the process, the number of phages to be purified and prepped for use, and the level of purity required to meet the FDA standards. Yes, it was possible. To do this for a critically ill person instead of the usual basic science experiments made it a little terrifying, but that didn't change the time required, and this was something he knew well how to do. This was his specialty. Jeremy had come from Australia for his postdoctoral fellowship a few years before, and that's how he'd come to be focused for so long on this particular phage purification process. His work largely done, he planned to head back to Australia in just a few months. But this was no time to wind down.

The SDSU lab was state-of-the-art, but it was an experimental facility—they had never prepared anything for human consumption, if you didn't count the food they brought in for departmental parties. The next morning, everyone in the lab gathered for a roundtable session to plan. Anca joined the tag-team effort to keep every step of the process on as tight a turnaround time as possible.

The purification protocol they had developed was a process that first concentrated and cleaned the phages, then soaked, chilled, and eventually centrifuged them to separate out the endotoxin, then tested again to be sure the phages were still active. Multiply that by four different phages, a snag in getting a needed test kit overnighted from Germany over the weekend, and additional steps needed to reduce stubborn endotoxins, and everyone was worried that the clock—and Tom's luck—would run out.

At each step, the phages checked out: highly active. They performed sterility tests and confirmed that no other microorganisms—bacteria or fungi—were present in the lysates. And finally, the moment of truth: the last-round endotoxin test. They fed the data from the previous steps into a spectrophotometer, which uses light to measure the density of endotoxin particles in the prep, pressed the Read button, and waited for the results to display on the screen. The initial numbers looked good, but were they good enough to use the phages on Tom? The only way to be more certain was to do additional analysis of the raw data and chart it. Thirty minutes later they had their answer.

They'd measured the endotoxin concentration of the starting phage lysate that the Texas lab had sent. It read 61,965 endotoxin units per milliliter. The FDA required the cleaned lysate to measure less than 1,000 to be safe and maintain a high phage dose. Finally, the cleaned phage results came back: 667 endotoxin units per milliliter. They had reduced the endotoxins by almost a hundredfold in less than twenty-four hours. The team high-fived.

Tuesday morning, Forest and Anca handed the vials of phages, bubble-wrapped and boxed for the drive, to Dr. Ji Sun in the investigational pharmacy at UCSD's Thornton Hospital. The phage cocktail was on the last leg of its journey, to the pharmacy and then its final stop: Tom.

Tom: Interlude VI

The desert is a vast sea of red sand, scattered with the remnants of invertebrate corpses. Mine will soon be among them. I am walking on a carpet of petrified fossils, flesh, fur, and feces, formed over millennia. My mouth is a dry crevice. The river that is my blood is thick, dancing along the artery, congealing. The transition from organic to inorganic, animate to inanimate, is ephemeral. We simply begin to exist as carbonized matter, as the water that comprises most of our bodies vaporizes. Our elements return to the earth, in a world where there is no more pain or suffering. I long for that world.

Although I cannot feel my legs, my feet shuffle onward toward an unknown destination. I think of my great-grandfather, who my relatives told me was a Cherokee who walked alone from Oklahoma to Texas along his own Trail of Tears. A profound sense of desolation and despair washes over me, like a gust of ancestral wind. My people are generations of untouchables.

I face a relentless, sunless sky with blazing heat. Night never falls, so the moonlight cannot guide me. I climb a dune. From this vantage point, I see what appears to be a copse of lush green trees in the distance. An oasis or a mirage? When I close my eyes, figments of my imagination are glass shards piercing my conjunctiva. My mind is a charlatan, playing tricks. The fever sings in mental wires. Do I even attempt to try to reach this alleged sanctuary, and risk having hope crushed by the heel of an unseen boot? Or do I clutch my apparitions like a shroud around me with gnarled hands, holding them as enemies, closer than friends? I long to be a part of civilization. To be human. Not a leper.

I am struck by the realization that if I am loneliness personified, then I must still be alive. If faith is knowing what I hope for and being certain of what I cannot see, then surely, I am lacking in faith. But I summon what little energy I have left and trudge on; my feet are a lifetime of burning in every moment. As I draw closer, I can see that the trees are genuine, and they are alive. But they are not lush; they are green on top and black with blight on the bottom, from faulty photosynthesis. A small pool of water surrounds them. Every cell in my body cries out for its nourishment. Its inky liquid has a slick, metallic sheen and appears thick, like an oil spill. But I will drink it. I must. I am dying of thirst.

Before I reach the water, three men materialize. They sit around the pool in lotus position. One is the holy man who presented Steff and me with the leaf. The other two are quiet-voiced elders with skin like bleached coral, as white as their robes, devoid of melanin. Their hair, too, is uniformly long and white. They watch me impassively. They have been expecting me. Just as suddenly, a blowfly appears and lands on the corner of my eye. It has been expecting me, too. In search of salt or a place to lay eggs, it flies into my mouth. I try to swallow it and choke. The sound startles the three men, who begin to talk in a foreign tongue, to each other and then to me. I know at once that they are deciding my fate. They ask me three questions, but I do not understand them.

The first-met stranger speaks to me in English, but his eyes do not meet mine. "You have not eaten enough of the leaf," he says admonishingly, "and you did not answer our questions. You must remain here for another seven years."

I hear another man's words, through a tunnel, from far away. "Failure to thrive . . . ," the hollow voice says.

The three men rise and leave, their long white robes flowing. At once, the pond transforms into a swirling cloud of blue-black, buzzing blowflies. As the swarm lifts and moves toward me in a black halo, the bones of a ram are revealed, picked clean. Its horns are diabolically twisted, and its open jaw gapes like a reaper. Their last host devoured, one thousand compound eyes are now upon me, their forelegs rubbing together as in prayer before a feast. Their mandibles exude digestive enzymes that will facilitate putrefaction. I want to welcome them to the task, but there are no tools to end my misery. In desperation, I clutch handfuls of sand and swallow them; the grit tears at my pharynx and esophagus. I look down to see sand pouring from my ribs, dripping as if from an hourglass, onto the desert floor. I fall to my knees, throw my head back, and howl, my spirit broken.

PART IV

The Darwinian Dance

Nothing is to be feared, only understood.

—*Marie Curie*

20

THE BLOOD ORANGE TREE

March 15, 2016

The next day, March 15, will remain burned in my memory forever. As I rinsed my coffee cup at the kitchen sink in the quiet light before sunrise, a pair of Bullock's orioles landed on top of the blood orange tree Tom had planted eight years ago. They were migrants that wintered in Mexico, flew north to breed in Southern California in the spring, and left by early fall. Tom and I tracked their movements regularly when they took up their annual residence in the palm trees behind our backyard. The male had bright yellow and black plumage that never failed to take our breath away. The female sometimes fed at our hummingbird feeder. This was the first day I had seen them this season. It felt like a sign.

I arrived at the hospital earlier than usual, around seven, and stepped out of the elevator to see a large group of doctors surrounding Tom's bed. I immediately panicked and thought the worst. *Had Tom died in the night?* Then I recognized a few of them; it was the nephrology team. Tom was still alive, but their visit could mean only one thing. They were ready to start dialysis on a second's notice. The senior nephrologist was a colleague I knew from a research project we had discussed a year ago, Dr. Joe Ix, chief of the nephrology division. After gowning up and entering Bed 11, I stood at Tom's side.

Joe looked at me in surprise. "Steffanie! It's great to see you," he greeted me warmly at first, but then I saw him blanch when the reality hit him. "But I hope you are here for professional reasons and not a personal one."

I put my hand on Tom's and looked up at my colleague with bloodshot eyes.

"Hi, Joe. This is my husband."

"That's what I was afraid of," he said quietly, and put his hand on my shoulder in sympathy. "I am so, so sorry. I can't imagine what you're going through." I knew from his tone that he thought Tom was going to die. By now, I realized, so did most of the docs. "I can see you have already met some of my residents," Joe continued. "We need you to sign this consent form for kidney dialysis. It could be hours or, at the latest, tomorrow morning, but I'm afraid it's imminent. We can't risk not being able to find you when the time comes."

I thanked Joe and signed the form with a wavering hand.

"With any luck, we'll be starting phage therapy sometime today," I told him. Joe had heard about our plans and wished me luck. He promised to call me personally when they began dialyzing Tom.

After the huddle of doctors left, I looked at Tom with newly sobered eyes. This might be one of the last times I would touch his skin while it was still warm. I ran my fingers over his face. Even through my blue gloves, I could feel the newly forming hollows around his eye orbits and under his cheekbones. His nurse had scribbled his weight on the whiteboard; he was now 187 pounds, which looked skeletal on his 6-foot-5-inch frame. He had lost over one hundred pounds.

Since Tom was feverish, we tried to keep him cool with ice packs under his armpits and a small fan. I had brought in a cooling sports towel that would remain cold for hours; I remembered buying it for him last summer when we had trekked the Virunga Mountains in Rwanda to see the mountain gorillas. A lifetime ago.

It felt like every random event was happening for a reason, beyond serendipity. Pandora started playing Madeleine Peyroux's version of a

Leonard Cohen classic, "Dance Me to the End of Love." Tom and I had seen her in concert only a few months before he fell ill. I sang the words to him and closed my eyes, heart aching in their irony, and not caring that I was off-key.

I laid my head on his chest as close as the snaking tubes would let me, knowing that it was the vent and not his own lungs that were responsible for its rise and fall. The sliding glass door to Bed 11 squeaked, and I turned to see Chip there, looking embarrassed that he was interrupting an intimate moment. I motioned for him to come inside.

Although Chip was unflappable, I could tell he had news; he was almost pulsating with excitement. "Jeremy just finished the endotoxin assay," he began. "The level is finally well below what the FDA required for safety." He waved a piece of paper like a flag. "So we have the eIND, and Jeremy is en route with Forest and Anca, bringing the phage prep back to the pharmacy."

I let out a whoop and looked down at Tom. Not even a flicker of movement. Chip added that the Navy had finished their endotoxin purification assay as well and were almost ready to ship their phages to us. And the FDA had made an unprecedented move, he said. In the past, they required that a separate eIND be issued for each drug—or in this case, each phage—but they had conferred and decided that an eIND could be issued for each phage cocktail. It sounds so simple, but it was a huge step in streamlining the process.

"At least we won't die from the papercuts incurred in the FDA's administrative bureaucracy," Chip quipped. "All joking aside, they have been incredibly helpful, and they are deeply interested in the outcome of Tom's case."

I was glad to hear that and explained to Chip what Joe Ix had just shared: kidney dialysis was imminent.

"Yes, I could see that's where things were headed," Chip replied. "But we can start phage therapy as soon as the pharmacy gets the phage cocktail titrated and prepares the buffer," he assured me.

Good news, at last. Mix, stir, and send upstairs to Tom—finally!

"So, if you're in agreement, the plan is to administer the Texas phages directly into the drains in Tom's abdominal cavity to reach those abscessed pockets," Chip explained. The drains would then be closed temporarily, trapping the phages and *A. baumannii* together so the phages could do their work.

"But I wanted to talk to you about the Navy phages, which are more potent. Carl advised that since Tom is now fully colonized with *Acinetobacter*, we should administer those phages intravenously."

Chip looked at me to gauge whether I understood the significance of his question. I did.

"I was afraid you were going to say that," I replied, with a huge sigh. "Ry warned that since his bacteria are multiplying every thirty minutes, any that are resistant will multiply and take over. So, if we're too conservative and only inject phage into his abdominal cavities, then we could miss reservoirs of bacteria that the drains can't reach."

"Precisely," Chip said, impressed, or maybe just relieved, that I was up to speed. "Of course, with IV phage administration we greatly increase the risk of septic shock. But if we start with intracavity administration and Tom tolerates it, then I feel confident about moving to IV administration as phase two. The benefits are worth the risk."

I didn't have to think twice. This was the best chance to save Tom. And if he died? Well, this would be the first case of IV phage therapy to treat a systemic superbug infection in the US. We would still have gotten important information that could help others, and he'd made it clear how important that was to him. If he were able, I knew he'd insist on it. *No retreat.*

"Let's go for it. We didn't come this far to hold back now."

Chip nodded. I got the impression that he had thought it was going to be a lot more difficult to convince me.

In the spirit of the day, I presented Chip with a little gift bag. He looked at me in surprise, and pulled out a little box of chocolates that I had regifted from a former student who was now based in Kentucky.

"Look closely at the box," I told him with a smirk.

"Bourbon balls!" he cried, and let out a chuckle.

"Yes, indeed, my Alabama friend," I told him. "Clearly you have a pair."

****

With Tom's kidneys now beginning to fail, his nurse, Chris, was recording his urine output on an hourly basis. I watched every drip of urine flow from his catheter into a plastic, flask-shaped bag, knowing that the moment it dipped below the threshold, the nephrology team would descend like a cloud of flies to start dialysis. The magic number was 30 milliliters per hour. If Tom made less, it meant his kidneys were functioning at less than 15 percent.

Every once in a while, Tom would grimace, his eyebrows would furrow, or he'd flail his hands in the air when he wasn't restrained, reacting to some invisible threat only he could see. Otherwise, he lay immobile. Comatose.

The wait for the cocktails to come up from the pharmacy was excruciating. Every clock in sight seemed frozen in time, as they had been in Luxor, waiting for the airlift out. The purified phages had been delivered to the investigational pharmacy hours ago, and I'd just assumed they'd be measured out into doses and be up shortly, but that wasn't happening. *Was there yet another snag?* I tried focusing on Tom and on practical details and always the music.

I swayed my hips to the endless music loops playing on Pandora, which was now Creedence Clearwater Revival's "Have You Ever Seen the Rain," and lifted the catheter tubing to usher the newly formed drops of urine into the bag. The color wasn't liquid gold, but a dark orangey-red, like a leftover jack-o'-lantern. I'd have to remember to razz Tom about this and give him the fascinating lecture on how to "read" the chemical composition of urine—the quantity, clarity, color, hue, fizz, and foam—for signs of kidney function and liver ailments, as well as inflammations, infections, diabetes, heart disease, and certain medication issues.

Admittedly, those signs weren't looking good for him just now, given the cloudy, darkening color palette of late and the Halloween hue in his catheter, but they were just barely good enough to hold off dialysis and hold on to the hope that the phages might change things for the better. Kidney failure is so critical to avoid because as they falter, waste products build up in the blood, the body's electrolyte balance goes off kilter, and organs start to die off, including the brain.

"Forty mils this hour," I reported dutifully to Chris in a soft voice, as he entered Tom's vitals into the computer. Blood pressure: 90/55, and that was supported by three pressors. Heart rate: 133, still tachy. Respiration rate: 29, despite maximum vent settings. And his latest creatinine level—the best measure of his kidney function—was 3.9. That meant he was deteriorating from AKI—acute kidney injury—to full-out kidney failure. So, heart, lungs, and kidneys were all on the verge of collapse. That was how the man in Bed 9 had died; one moment, it was a bad case of the flu, the next, his heart gave out.

Boom. Just like that.

Since epidemiologists like me crunch numbers for a living, I usually take solace in statistics. But of the 4 million admissions to ICUs across the United States each year, up to one in five patients die. I was praying that Tom wouldn't be one of them. He had been back in the TICU for nearly two months, way longer than the three-day average, and this was his second time there. Although they rotated in shifts, by now the girls and I knew the names of most of the nurses and doctors, physical therapists, lift team, and housekeeping staff. And they knew ours. We were family of sorts now, comfortable enough to be candid, candid enough to be irritating. Me, especially, with my habit of hyperfocus and constant questions. Sometimes, I'd hear myself pressuring for answers no one had, but I couldn't stop.

I looked at my watch. Mid-afternoon. Just when I was getting ready to rattle a cage or two, Chip appeared at the doorway to Tom's room, Bed 11. His role as the protocol chair for Tom's phage therapy meant that he was the point person for the entire operation. I rarely saw him without a smile. Today, the freckles stood out more prominently on his pale cheeks. He was nervous.

"Chip, what in God's name is taking the pharmacists so long?" I whined, wringing my hands.

"I know, the wait is gut-wrenching for me, too," Chip admitted. "But the research pharmacy is used to preparing medications for clinical trials, not viruses. Think about it: this is the likes of which they have never seen. And they not only need to get the dilutions right, but each bag of phage prep needs to be dispensed with the right volume—a billion phages per dose—then labeled with its contents and the eIND number. The techs also need to make up a buffer solution so that the phages enter an environment with a neutral pH. And we need enough to dose him every two hours, at least for now."

Okay. I remembered this from my reading and Ry's phone lecture. Since we would be injecting the Texas phage prep into the drains in Tom's abdominal cavity, the phages could face an acidic pH from the bile, ascites, and stomach acid, so a buffer solution of water and sodium bicarbonate was being prepared to neutralize their surroundings. Got acid? Get a base, and mix them to neutralize it. High school chemistry. Alka-Seltzer without the *plop* and the *fizz*.

Suddenly I realized that there was a key detail to the protocol that I had overlooked. "How did you decide what concentration of phage to have the pharmacy prepare?" I asked Chip quickly, while I combed the library in my mind. I couldn't remember reading a single article that advised on this topic. Chip fidgeted with a piece of paper that he took out of an envelope as he thought about how to respond. He was in a tough spot. I was not just his friend and colleague, but the wife of his patient, Patient Zero in an unprecedented experiment that the whole phage community, Navy research lab, and FDA was watching. Nah, no pressure.

"Carl and Maia were very helpful," Chip replied. "But basically, no one knows the ideal concentration of phages to optimize their killing potential while minimizing the chance of septic shock. And when it comes to IV phage administration, it's been done in animal models, but there are no documented cases in the literature. That is, in humans."

Chip looked up from the paper he now held in his hand, and his eyes met mine with an even gaze. "We're using standard IV dosing measures that take into account Tom's weight and the endotoxin level, as well as the potency of the phage cocktail. So, we can estimate pretty closely how much phage we are injecting into Tom's drains or his bloodstream. But the phages will multiply to an unknown extent. It's a double-edged sword; we *want* them to multiply to eradicate the *Acinetobacter*, but we can't anticipate *exactly* what they will do once they are inside him. It would be easy to think about it as warfare, in which each side takes on new weapons, but in reality, it is survival of the fittest. There's a complex interplay between the phages' antibacterial activity, Tom's immune system, and how quickly the surviving bacteria take the lead and proliferate. We just don't know enough to predict how that will play out. To be perfectly honest, we are flying by the seat of our pants here."

I stood there nodding, but felt numb inside. This was a classic catch-22. Do nothing, he dies. Do something, he might die anyway.

Fuck.

"So, you are saying that it's a guess?" I asked Chip, looking him directly in the eye. It was Sir Isaac Newton who reputedly said, "No great discovery was ever made without a bold guess." We were about to test his hypothesis. I remembered that the doctors who treated Anne Miller, the first person to be treated with penicillin in the US, had not known what dose to use either. *No, I am not alone.*

Chip took a deep breath and thrust the paper he was holding toward me. "I need you to review the informed consent form for the phage therapy. Let me know if you have any concerns or questions, and if you agree, sign it," he said.

Consent to Emergently Administer Bacteriophages in the Treatment of Multi-Drug-Resistant *Acinetobacter baumannii* Infection

I understand that my husband has a life-threatening infection with a multi-drug-resistant *Acinetobacter baumannii* strain or strains and that, despite efforts to treat the infection with antibiotics and drainage, he remains critically ill.

I have expressed an interest in exploring whether experimental treatment of his infection with bacteriophages can be directed at my husband's organism. I understand that bacteriophages are best described as "viruses that attack bacteria" and that there has been some experience treating human and animal infections with these organisms under experimental conditions but that these agents are not approved for clinical use in the United States or Western Europe. I understand that physicians and scientists at UCSD, Texas A&M University, AmpliPhi, and other research laboratories have collaborated to identify phages that have activity against my husband's organism in laboratory-based studies. I understand that although great care was made to perform these studies and to prepare phages that might be used to treat my husband as safely as possible, these efforts have been done under a very short time frame using approaches and laboratory reagents that are used for research studies, but some of these approaches and reagents are not ones that would be used for the preparation of clinically used medicines. Thus, the laboratories, biotechnology companies, physicians, or scientists who have participated in this effort cannot provide assurances about either benefits from or the safety of the approach.

The potential benefits and side effects of bacteriophage therapy for bacterial infections have not been extensively studied in human clinical trials. Although it is possible that these bacteriophages will decrease the amount of *Acinetobacter baumannii* in my husband's abdomen or other areas, there are no assurances that this will happen. Because of the limited experience using bacteriophages to treat bacterial infections in man, all of the potential side effects of this study cannot be predicted at this time. It is possible that administration of these materials could result in a worsening of his condition or even his death. It is possible that my husband could have an adverse reaction to the bacteriophages or to other substances (including bacterial endotoxin or other substances) that might be present in the material. These side effects could include a reduction in blood pressure, changes in his heart rate, and/or damages to organs including (but not limited to) the lungs, liver, and kidneys.

If these bacteriophages are used in an effort to treat my husband's infection, I understand that one or more of the bacteriophages that have been demonstrated to have activity against my husband's organism in the laboratory will be administered through catheters that have been previously placed in an effort to drain tissue collections of the *Acinetobacter* within my husband's abdomen. Depending on his condition and his tolerance of the instillation of the bacteriophages into the abscess cavities, I understand that one or more of the bacteriophages might also be given by intravenous, intraperitoneal, or oral routes.

I understand that those caring for my husband will closely follow his clinical condition during and after the bacteriophage administration and that they will perform laboratory studies designed to determine whether the bacteriophages have decreased the amount or changed the nature of the *Acinetobacter baumannii* (or other bacteria) that are cultured from my husband. These studies will include studies performed in the UCSD Medical Center Clinical Microbiology Laboratory but may also include specialized studies performed in research laboratories at UCSD, Texas A&M, or other research entities.

I have been provided with a copy of this form and given an opportunity to ask any questions that I might have. I have been told that I can ask additional questions at any time and that I can reach Drs. Schooley or Taplitz through the UCSD Medical Center or on their cellular telephones. I understand that the bacteriophages are being administered under an emergency, single-patient Investigational New Drug Application [Individual Patient, Emergency 21 CFR 312.310(d)] issued by the US Food and Drug Administration and that, as per FDA guidance, the UCSD Human Research Protections Program will be notified of this intervention after the fact.

Understanding that my husband's clinical condition is grave and that there is a substantial chance of morbidity and mortality with or without the experimental use of these bacteriophages, I consent to their administration on behalf of my husband, who is not currently sufficiently alert and oriented to provide consent for himself. I understand that the alternative to proceeding with this research study is to continue medical therapy according to clinical judgment and recommendations of my husband's treatment team. I understand that I can withdraw my consent at any time for any reason and that a decision about whether to proceed with bacteriophage therapy or to terminate bacteriophage therapy will not prejudice my husband's care by his medical care team in any way.

Steffanie Strathdee, PhD *Date*
On behalf of Thomas Patterson, PhD

Witness *Date*

As I read the consent form over, my stomach did a somersault. I had seen an earlier version of this form a few days ago when Chip was preparing the necessary documents to obtain FDA approval for the phage therapy, so the contents shouldn't have come as a surprise. But at the time, the Scientist Me had read it. This time, I was reading it as Tom's wife, signing it on his behalf as his power of attorney.

A new reality was setting in. If Tom died of septic shock now, it would be because this was a treatment that *I* had sought out and initiated, albeit with the help of Chip and others. What if I killed him? How could his daughters ever forgive me? How could I ever forgive *myself*? The immense responsibility of this dilemma was suffocating. Surreal. Just three months earlier, Tom and I had been exploring a pyramid, having the time of our lives. Now he was in a coma and we were about to inject viruses into his body to save his life from a superbug. *I can't believe this is our life*, I thought miserably. Tom always said that we had horseshoes up our butt. We sure as hell needed them today, along with some phages that could swim upstream.

As I signed the form and handed it back to Chip, my eyes were gleaming with a combination of tears, fear, and hope. "I don't know why, or how to explain it," I said forcefully, "but I think this is going to work." There was absolutely no concrete evidence for me to think this way. Call it instinct, a gut feeling, or a spidey-sense that had rubbed off from my sessions with Robert, who remained convinced that it wasn't Tom's time.

"I do, too," said Chip, with a tight smile. "I do, too."

I remembered what the scientist and Nobel Prize winner Barbara McClintock had said in 1983 after receiving the award for her discovery of genetic transposition, which showed that genes could "jump" and switch the physical characteristics of cells on and off. She had done most of her research in the 1940s and '50s, but had been on the margins of the scientific community for decades. She was brilliant *and* she trusted her instincts. "If you know you are on the right track, if you have this inner knowledge, then nobody can turn you off...no matter what they say." Hell, yeah, sister.

We didn't have the luxury of time for debate any longer anyway. Carly had mentioned that one of the medical residents had pulled her aside the day before and said, "If something doesn't work soon, we're going to have to start unplugging things." We'd been ever hopeful, but the truth was that Tom would die if we didn't do something drastic. We had nothing to lose.

Chip left for a while to check on his other patients, and I watched Chris empty the ostomy bags attached to the five drains in Tom's abdomen. These were filling up every few hours with a thick, cloudy, purulent liquid that ranged in color from yellowish to brownish; a mixture of pus, ascites fluid, bile, and god knows what. There are no medical terms that describe its smell: like a swamp. I knew from the last round of bacterial cultures that the ooze from at least three of these orifices was teeming with *Acinetobacter*, which had crowded out most of his normal gut flora.

"Hey, don't throw all that out," I called over to Chris as he emptied the bags. "We need some for the baseline samples." He raised his eyebrows at me and wrinkled his nose, as if to say, *Seriously? Shouldn't you be focused on what might be the last few hours you have with your husband?* But Chris was calm, professional, and polite enough not to say this out loud. Or maybe he wasn't thinking that at all, and it was my own inner voice, scolding and not at all polite.

I cleared my throat and explained my rationale. "I know it seems crazy, but aside from being husband and wife, at our core, Tom and I are both scientists." I stopped and took a deep, shaky breath, realizing that I was also justifying my detached demeanor to that other part of myself. "Even if Tom dies, we need to learn something from the phage treatment. And if it works, we need to document it as carefully as possible so it can ultimately help others. If we forgot to do this and Tom were conscious, his dying words would be: 'What do you mean, you *forgot* the baseline sample? What kind of researcher are you?'" I laughed, and it sounded a little hysterical. "So please, no matter what, save everything you can."

Chris nodded. He'd been an ICU nurse for a while, and although this was probably not something most people brought up in moments like these with their loved ones, he knew this case was being documented as medical research, and as such, data collection had to be just so.

At some point, as the afternoon wore on and predictions about when the phages would arrive grew later, it was clear that my constant calls and texts to everyone for updates weren't actually helping Tom or anyone else. I was only adding to the burden for those who were meticulously monitoring Tom and, in the pharmacy, carefully diluting the phage cocktail to prepare them for injection.

Tom, however, was a captive audience, so I just kept nattering to him about every possible thing—phages, orioles, Newton, and the kittens—as if my talking to him kept him grounded in this world. With me. If Tom could hear me through the coma he was probably ready to jump out of his skin just to escape. I could see myself, from outside myself, in this manic mode, and I saw a desperate woman, but I couldn't get her to calm down.

Just then, I saw a familiar figure with a red bobbed hairdo make its way down the hallway pulling a rolling suitcase. My friend Michelle from Vancouver was a flight attendant and had managed to snag the last vacant seat on the daily nonstop flight so that she could come and rescue me. By the look on her face as she peered into Bed 11, I needed it as much as Tom.

"Hey, girlfriend," Michelle began, looking past me at Tom. I could tell she was shocked.

"Don't come in without gloves and a gown!" I snapped at her, stunned at how harsh my voice sounded. I hadn't even said hello.

"Whoa," she replied, putting her hands up in defense. She pulled a yellow gown over her head and snapped on blue gloves. "Sounds like you need some R&R, big time."

My eyes filled with tears for what must have been the millionth time.

"I'm so sorry," I told her, shaking my head in disgust at myself. Like everyone around me, she was doing the best she could to help. I caught Michelle up on the plan for the day, and the long delays we had faced.

She crept closer to Tom and patted his hand. "Hey, Tom," she said. "It's Michelle, if you can hear me..."

She turned to me. "I wouldn't even recognize him," she said. "He looks, he looks..." She didn't finish her sentence, but we'd known each other for twenty years, watched each other raise kids, and had had our share of laughter and tears together. *Like a corpse*, her face said. She turned back to me.

"When did you eat last?" she asked.

I had to think about it. "I had a banana on the drive in this morning."

Marilyn came up behind me, standing at the doorway, and introduced herself to Michelle. "I've got an idea," she said brightly. "How about Michelle and you take off and have some time for yourselves. Carly and Frances are both due in this afternoon, right? If it looks like the phage therapy is going to happen, one of us will call you and you can come back."

Marilyn looked at me expectantly. Her haircut and calm demeanor reminded us both of our friend Heather back in Vancouver. I looked at Michelle, then at Tom, then slowly nodded. She was right. I was doing no one any good at the moment. Frances would be here any minute for her afternoon shift. Carly, too. Tom wouldn't be alone. And I needed to summon my own kind of mindfulness if I was going to get through the next few days.

I asked them both to give me a few minutes with Tom, so I could kiss him goodbye. How many times had I done this, not knowing whether it would be the last time? Too many to count.

"You know I love you, and I am with you always," I whispered in his ear as I stroked his cheek. "I'm gonna take a break while we wait for our phagey friends, and I'll see you later."

Okay.

If Tom's spirit was journeying, as the girls had suggested, I just prayed he could find his way back. He was still motionless as a mummy in his shroud of a gown, interwoven with IV lines, drain tubes, patches, and sensors. No perceptible *ka*, and he'd need it to fight off the enemy within. I thought about Khalid and what a patient teacher he'd been, and a life-saving guide far beyond the pyramids, when we'd depended on him to help us navigate Tom's medical crisis in Luxor. What was that myth he'd told us about Ra, the sun god, descending into the underworld, battling demons and gods who opposed him? Apophys, the god of chaos, would partially swallow Ra, leading to the setting of the sun each night, and then would spit him back out at dawn, for the sunrise. At this point, tonight's sunset could easily be Tom's last. He needed to hold on for another dawn.

21
MOMENT OF TRUTH

March 15–16, 2016

When the call came at around seven p.m., I nearly fell off my chair. Michelle and I were sipping our second glass of Chardonnay on the back patio, watching the sunset, and trying to keep our minds off the inevitable. It was Tom's favorite time of day: dusk. Not quite light, and not quite dark, like the in-between netherworld he was suspended in. I'd fed the kitties and reheated some tamales that one of our fellows, Argentina, had dropped off earlier in the week. The girls and I had tried to share cooking duties, but the days were exhausting and sometimes Tom's bad turns meant that we stayed longer than we planned. So, when friends and our students offered to cook for us, we gratefully accepted. One of our postdocs organized a schedule, and people would drop off casseroles at the hospital or our house. Nepali and Indian curries, Filipino spring rolls, Mexican tamales—a world of foods and such generosity, comfort food in the truest sense.

I poured that second glass of wine for us both. Nothing was going to take the edge off this day, but it was time to at least hit pause on the caffeine. My cell phone trilled, and I answered with a bark that shattered the peaceful quiet. Michelle winced. It was Carly.

"It's time," she said excitedly. "The phages are finally ready!"

She quickly brought me up to speed on what had transpired over the last few hours. For his part, Tom had kept peeing, which meant kidney dialysis was again deferred until the next morning. Whether Tom could last that long was anybody's guess.

It had been Chip's turn to get antsy; he had called down to the pharmacy every hour as the late afternoon ticked by. Chip had once told me that he didn't like patients to have procedures in the evenings because there were fewer staff around, which is why I had started to suspect that phage therapy wouldn't be started until morning. But by the sound of things, Chip had overruled himself. He told me later that he was too worried that Tom wouldn't make it until morning. There really wasn't a minute to lose.

"So," Carly said, "do you want us to wait so you can get here?"

It was about seven o'clock, and the view of Highway 5 from our backyard still showed a trail of red lights running south as rush hour dragged on. *Bloody hell.* How could I have worked so hard to make this moment happen and now miss it? I was crushed that I wasn't there, but delaying the phage therapy was not an option. It was go time. Carly and Frances would keep texting me so I could follow the play-by-play. If things started heading south, I would call an Uber and get there ASAP.

In a sudden streak across the weary sky, a peregrine rocketed by and perched in the palm tree in the neighbor's yard, scoping out the back slope for pigeons and ground squirrels. Tom would have appreciated the falcon's familiar presence at his favorite time of day. I was struck by a harsher truth that was oddly reassuring as well. For every prey, there is a predator. In the peregrine's case, with the ability to fly as fast as two hundred miles per hour, it is the perfect predator. Now we were waiting for another perfect predator to do its work.

Over the next few hours, Bed 11 became the busiest room in the ICU, but in a good way for once. At least ten doctors appeared seemingly out of nowhere, representing a full range of medical disciplines: infectious disease, pulmonary medicine, GI, nephrology, cardiology, and interventional radiology. In addition to those on duty, some TICU staff and a clutch of residents and fellows who were on rotation hung out at the door when they were not needed elsewhere. Cara Fiore, the FDA official who had coordinated the approval process, had called Chip from her son's hockey game in Maryland to check on the patient. No one wanted to miss being a part of the action. History was being made, no matter what the outcome—and no matter what time it would finally happen.

Frances and Carly were at their dad's side, taking turns holding Tom's hands. Martin, our holistic healer, had arrived earlier and gave each of them a hug. He had known them both since they were teeny-boppers. From his home in Toronto, Robert texted to tell me he was "tuning in" by making a psychic connection to Tom. Around the world, countless numbers of our friends and colleagues were praying, lighting candles, sending energy, and checking Facebook for updates. I later learned some had been avoiding checking Facebook in recent days because it seemed inevitable that Tom was dying and they didn't want to see the dreaded post that would make it real.

At a little after seven, Dr. Sun, the manager of the investigational pharmacy's drug service, and his resident, Minh, finally appeared at Tom's doorway, carrying a large foam box that had a biohazard sticker emblazoned on top. An unassuming man, Dr. Sun was not used to being the center of attention. He blushed when several of those in the room gave a quiet cheer, greeting him with the enthusiasm that one might give the ring bearer at a royal wedding. He stepped toward Chip and, with a light ceremonial bow, handed him the box.

Chip pored over its contents, checking that each phage was correctly labeled on the exterior of a vacuum-sealed plastic bag, which had been carefully kept at four degrees Celsius. The plastic was dark brown to protect the phages from the light, since they are sensitive to UV rays. Randy counted out one syringe for each of Tom's three drains that had tested positive for *Acinetobacter*, as well as three syringes for the buffer solution.

"We need Martin to bless the phages!" Frances cried out suddenly. Carly nodded in agreement. Martin stepped forward, approached the open box, and laid both of his huge, gloved hands on the brown plastic bag of phage prep. His figure was imposing; he was almost as tall as Tom. He closed his eyes and muttered an unintelligible prayer. Several of the physicians bowed their heads and murmured an awkward "Amen," while others gaped or looked away. The room felt charged with electricity, all focused on the phages now, as if every molecule of energy focused on Tom's recovery had been harnessed into a laser beam.

Randy handed the syringes of buffer and phage to Dr. Picel, the IR doc who had personally inserted all of Tom's drains and his feeding tube. He literally knew Tom inside and out. Just then, Carly whipped out her smartphone and snapped a photo of Chip and Randy at bedside, smiling nervously for the camera. Then she blasted the room with the chorus from Survivor's song "Moment of Truth." Although this was by far the scariest moment Carly and Frances had likely ever experienced, I marveled that they welcomed it with their dad's offbeat sense of humor. He would have roared with laughter, but his spirit was somewhere else, deep in the ether. As everyone in the room inhaled, Dr. Picel injected the buffer and then the phages into each of the three drains. The silence was punctuated only by the muted beeps and wheezes of Tom's monitors and the infernal drip of the ventilator hose, until Frances asked the question everyone was thinking.

"What happens now?"

"We wait," Chip responded. "We wait, and hope that the next twenty-four hours are the most boring we have ever had."

That would depend wholly on how boring the next twenty-four hours were for the phage and *Acinetobacter* as they squared off in their own fight for survival—not against Tom but against one another. The scientific literature and high-tech electron micrographs suggested anything but a boring encounter.

••••••••

Phages don't so much prowl for their prey as bump and sniff what they encounter to find a matching host bacteria. So it can take them a while to find a match if the concentration of their prey is relatively low. In Tom's case, his massive infection was a phage's smorgasbord. *Acinetobacter* was everywhere. An all-you-can-eat buffet. It was especially concentrated in the pseudocyst and nearby pus-filled abscesses where some of the drains had been placed, so that's where the first phages were injected for a direct hit. But the micro lab had also cultured it from his lungs and periodically from his blood.

In a typical phage strike, once the phage locates its target bacteria, it moves in for the kill. A recent study even found that phages cooperate in their attack, with the first engaging in a suicide mission that weakens the bacteria's defenses for the ensuing attack from the rest. But once the phages have put a dent in their target bacteria host population, bacterial mutants of *A. baumannii* that weren't vulnerable to the cocktail then proliferate. Those surviving bacteria are, by definition, resistant to the cocktail in use and thus pose a threat of reviving an overwhelming infection. The easiest way for bacteria like *A. baumannii* to become resistant is to simply delete the gene that encodes for the phage's receptor. With the receptor modified, the phage can no longer attach to its host.

Tom's *A. baumannii* was one of the nimblest superbugs on the planet, top-ranked not only because it picked up so many antibiotic resistance genes, like Tom's had, but because of its overall hardiness. Extreme heat, cold, or other environmental assaults? No problem, *A. baumannii* has evolved to survive. Chemical disinfectants and other sterilizing efforts to clean medical equipment and hospital settings? *A. baumannii*'s biofilms are impervious to some of them. Plus, it has evolved what have been described as "tiny fingers" to hold on to those smooth, slippery surfaces. Tom's *Acinetobacter* had also clearly beat out competing bacteria in Tom's microbiome. Unlike the mild-mannered *A. baumannii* I'd first encountered decades ago as a student, this one wasn't wimpy anymore.

But the phages didn't need to be superstars to score. With the element of surprise on its side, this first-round draft of phages, injected directly to the infection sites, could deliver a blow to the vulnerable *A. baumannii* that could maybe knock the infection back just enough for repeated rounds of phage reinforcements to break its grip and give Tom's immune system a fighting chance.

Invisible to us but invincible, we hoped, the phages were already now in pitched battle against the *A. baumannii*. In this theater of war where the action changed in a matter of minutes or hours, our field reports from the trenches, our only intelligence from the frontlines, would come in what suddenly seemed a primitive mode of communication: blood tests and analysis of the "output"—the gunk from Tom's drains—and from lab cultures and the digital displays on the monitors that crowded Tom's bedside, tracking his vitals. The generals in our war room—Chip foremost—had to figure out how to stay one step ahead of the enemy, working off information that was already hours old by the time it was decoded.

With the arrival of the phages and the start of treatment, it was tempting to imagine that the tide was turning and Tom might pull through. But when I closed my eyes to try a visualization practice Robert had suggested, one in which waves of phages crushed the *A. baumannii*, the image stuck in my mind was that of a killer wave about to crush my surfer dude.

••••••••

Through the night, Tom received round after round of the Texas phage cocktail, one every two hours, and there were no signs of adverse reactions. I stayed home but might as well have gone back to the hospital, as much as I called in to the TICU nurses' station to talk to his night nurse or anyone with eyes on Tom. Finally, well before sunrise, I took off for the hospital and was back by Tom's side.

As part of the protocol, samples from Tom's drains were to be taken every twenty-four to forty-eight hours and sent to the CPT and the Navy labs. For each round, the *A. baumannii* first had to be isolated and cultured by the micro lab, then sent along to the Navy and Texas labs. On their end, the labs would test to see if the phages in use were still active against successive isolates of *A. baumannii*.

I peered at Tom closely, studying him for some sign, anything, that would reveal if the phages were working. But I could see no change in his appearance. His labs looked the same as the day before, with the exception that his white blood cell count—the blood's infection fighters—had increased and his hemoglobin had dropped, so he was given yet another blood transfusion. Chip explained that this was expected; Tom's immune system had detected an invader. We could only hope that this was a mild episode of friendly fire.

Tom's vital signs remained relatively stable, although "relatively" at this point was nothing like "good." He remained in a coma. And he needed every bit of life support to breathe and keep him this side of a flat line.

Joe Ix, my nephrologist colleague, stopped in to check in on Tom's progress. Tom's hourly urine output had increased somewhat, to 60 milliliters per hour.

"Your husband pees like a racehorse," he grinned. High praise from a nephrologist.

"Yeah, and by now, he probably costs more than one, too," I replied with a tired smile. So far, so good. Joe decided that Tom had avoided dialysis for yet another day, and he would check in again tomorrow.

The Navy phages were scheduled to arrive by evening, with the pharmacy team on standby to prepare them for IV administration. As Chip had warned, this was infinitely more dangerous due to the potential for septic shock. I knew how fast that could happen, since I had seen for myself how suddenly Tom had gone into shock when his pseudocyst drain had slipped two months ago.

We were entering even more uncharted territory now. Over the next few days, Tom's life would hang in the balance. We all knew that one of two things would happen soon.

He would recover, or he would die.

22

THE BOLD GUESS

March 17–18, 2016

For all the hoopla, the do-or-die moment of injecting phages into Tom's bloodstream seemed oddly anticlimactic. I half expected him to open his eyes or to sit bolt upright, but there was nothing of the kind. Of course, that was unrealistic to expect. That's how it would play out on TV. Tom had been comatose off and on for nearly two months before the Texas cocktail had been injected into his abdominal catheters Tuesday night. Now, forty-eight hours later, and with much less fanfare, Randy Taplitz administered the first dose of the Navy IV phage cocktail to Tom. We both held our breath as the phage dilution surged through the IV line embedded in Tom's frail arm.

We both knew this was the riskiest moment thus far, and silly as I felt doing it, when no one was looking I took a cue from Frances and blessed the phages—just put my hand on the bag and said a little prayer. *It couldn't hurt.* While I prayed, I imagined the phages swimming to the infected abscesses, evading whatever Tom's beleaguered immune system might throw at them, and dodging the liver and spleen's filtering system that Carl's work had shown to clear phages from the bloodstream. Cocktail notwithstanding, it was not your average happy hour. Tom would have joked, *Make mine shaken, not stirred!* In my mind's eye, I lifted my imaginary glass and made a toast. *Here's to your health, Hon—please please live long and prosper.*

The procedure itself was remarkably simple—so simple that Randy called the pharmacy to double check, then taped the instructions to the wall for the rest of the team. I watched her press the plunger on the IV syringe, and thought of the microscopic "living syringes" flooding in to inject themselves into the *Acinetobacter.* Tom would have laughed—he'd been on the end of an experimental syringe before. He'd been just a kid in San Diego at the time Jonas Salk was conducting his polio vaccine research, and Tom was one of the schoolkids who received the experimental polio vaccine. He'd told the story of being in third grade, lining up for the shots in the school auditorium. The needles were huge, at least to an eight-year-old's eyes, and full of pink fluid. This was before they had perfected the polio vaccine, and while the study was successful in proving it could work, there were some adverse side effects for some children. Tom had come through that well enough, and the polio vaccine had gone on to nearly wipe the disease off the planet as a threat to generations since.

Whatever came of Tom's current encounter with an experimental treatment on the pointy end of a syringe, I just wanted him to live to tell the tale. That story would have to wait, and it might not be Tom's to tell if the phage therapy didn't work, or didn't work fast enough.

As ubiquitous as phages are in the environment and in our bodies, from birth to beyond the grave, and as integral as they are in the various microbiomes, surprisingly little is known about how they operate in our bodies. It was easy to think of phages simply as the good guys, predators policing our innards, picking off bad-guy bacteria like *Acinetobacter* as just part of a day's work. That's the mission we'd assigned to *these* phages. But the selection process for our phage cocktails had been so painstaking precisely because the gazillions of other phage types—an estimated 30 billion or so that we absorb into our body every day—include vast numbers with no interest in *Acinetobacter*, or at least not in Tom's particular strain.

All Tom's team could do was construct a cocktail of phages with promising characteristics, but no one was sure how they would perform *in vivo* on this war front. Or if they'd be able to survive long enough to meet the foe.

Bacteria don't have brains like higher forms of life, but it sure felt like Tom's *Acinetobacter* was plotting against our every move. If Tom were here—and by that, I mean really *here*—he would scoff at my anthropomorphism. After all, bacteria are just doing what they do to survive. We had relegated them to the bottom rung of the food chain, but they were showing us how shortsighted we'd been.

Despite modern-day advances in technology, one sign of success on a Petri plate hadn't changed since Félix d'Hérelle's day. You wanted to see the telltale sign that the phage was killing off the bacteria around it. Those Swiss-cheese-like plaques with clear zones around them that look like halos. We were all praying for halos.

••••••••

The challenge ahead would be twofold. First, to track changes in Tom's bacterial isolate, indicating resistance mutations by the *A. baumannii*, and then, if needed, to tweak the phage cocktails in strategic ways that could keep the phages' selective pressure on the bacteria. This was painstaking lab work, and it took time. Staff at both labs were still working around the clock, eschewing the usual distinction between work and after hours.

As the TICU team quickly got more comfortable with the phage therapy protocol, the nurses began handling Tom's phage cocktail infusions, administering the phage preparations into Tom's abdominal drains and his IV line, with the dosing schedule now reduced to twice a day. With no immediate window on the action between the phages and their prey, all we could do was watch Tom and the periodic lab reports for any indication of change, for better or for worse. So far, so good. Or at least an absence of bad.

Randy wrote in her notes: "Has had no immediate adverse reaction to infusions." Hurry up and wait was the order of the day. Days. Tom had been battling this bug for just over one hundred of them thus far, and counting.

23

LYSIS TO KILL

March 18, 2016

At the center of the low pulsing beeps and lights of the bedside monitors, Tom lay motionless and silent, as still and wasted away as a human being can be on this side of death. His gaunt, pale body looked like an abandoned battlefield, post-apocalypse. But in truth, the battle had just begun. The Texas phage cocktail infused through Tom's abdominal drains was barely four days along, followed by the Navy's IV phage twenty-four hours ago. Now we watched and waited for the first lab reports that would give us a glimpse of how the phages were faring.

Chip called it the Darwinian dance, but that's a lyrical way to describe a ground war that is more scorched earth and a battle to the death, all of it in a microbial universe. I'd like to say, "in a galaxy far, far away," but it wasn't. It was right here, ravaging Tom's body from the inside out. Tom was the battleground, the field of action where the *Acinetobacter* and phages were staging this Armageddon, a tactical war between them with stealth strategies and advanced genetic weaponry. In a sense, from their microbial point of view, this isn't even about Tom. This is one-on-one a battle between virus and bacteria, each one purely focused on its own survival.

* * *

Since the girls had returned, I'd spent the mornings through lunchtime with Tom, and they'd taken back-to-back shifts with him in the afternoon, sometimes with Danny. We all had our own routines, but at night, we reconvened at the house and talked over wine, often too tired to do more than stare like zombies at the TV, which was typically endless episodes of *Forensic Files*. Tonight, not even the TV offered a way out. We felt trapped in our own *Twilight Zone*. Our desperate hope that this novel phage therapy treatment would work was matched only by the fear that it would fail.

It was easy to feel untethered from time and reality. Now and again, someone would say something that would yank us back and down. Many of the medical concerns we raised must have sounded piteously naïve to anyone looking at Tom's situation from the outside. A few weeks earlier, we'd expressed concern about the continued use of opiates for pain management, since we didn't want Tom to become addicted. When we mentioned this to Meghan, always a straight-shooter with us, she'd shaken her head and said, "That's the least of our concerns right now."

Then there was the conversation I'd had with Randy that afternoon. I had a hunch I knew the answer, but I'd asked anyway. It was important to Tom.

"If Tom dies, can any of his organs be used for donation?"

She'd looked at me like I had two heads.

"No," she said. "Not with this kind of infection. Too dangerous."

"Even his corneas?"

"Even those," she replied.

That bastard bacterium. *A. baumannii* was calling the shots, and even if he died it was going to rob him of his one last wish—that his death contribute in some way through organ donation. I thought of how strong and robust a man Tom had been a few short months ago, and now every last part of him was dying and would be discarded.

The night before, we'd kindled a low flame in the firepit at twilight. Carly and Frances lamented again that we might not get a chance to hear his voice, so I pulled out my phone.

"I saved some voice mails," I confessed, and the girls both chimed in that they had, too. So, we all sat and played them on our phones, one by one.

Carly had one that she treasured that wasn't like the goofy messages her dad often left for her, saying something ridiculous in a funny accent. It was just an ordinary message letting her know we'd just returned from a trip and to give us a call when she could. She loved that it was a "normal Dad" message. We were all aching to hear that voice again.

Listening, I was shocked at how full of life his voice had been just a few months before, and now we were lucky to get a raised eyebrow or a hand squeeze. I'd have given anything now to hear Tom utter the most ordinary, ordinary, ordinary things again. Or just to believe in this moment that I'd ever hear his voice again.

I'd told Chip that I had a feeling the phages would work, and I'd meant it when I said it. But when the dark had closed in on the day, it was as if the setting sun had pulled my spirit, and Tom's, down with it to the underworld where the demons waited: Pain. Frustration. Grief. Guilt. The knowledge that this intervention was my idea—my fault—was becoming too much to bear. And fear of what lay ahead. Robert had told me years ago, after Steve had died and I was struggling with that loss, that in spiritual terms, I had experienced every major challenge and had acquired every skill I needed to manage whatever problems I faced ahead. "If you can see that any new challenge you face is just a variation on the ones you've overcome, you will succeed—and become a more enlightened being in the process," he'd said. Right now, I was anything but enlightened. In the dark, I broke down in tears, feeling the fool for ever having thought this scheme stood a chance.

****

Twenty-four hours later, I felt foolish for ever having doubted.

Carly had gone down for her usual afternoon shift at one p.m., but she was gone longer than usual. I was napping on the couch with Bonita, and Frances was in her bedroom meditating, when Carly burst through the front door bubbling with excitement. I had to do a double-take, it had been so long since anyone effused like this. Tom had come to—awakened from his coma! He'd been groggy, still intubated, so unable to speak. But he'd raised his head off the pillow, kissed her hand, nodded to Danny standing by her, then, exhausted, drifted off to sleep—just sleep, Carly insisted, not a coma. Although only for a few minutes, Tom had awakened. *He was back.*

The ICU experience, wherever you are in it, has a way of leveling all drama—clinical or otherwise—probably to keep everything at a manageable hum. The sounds, the temperature, the conversational tones—everything is calm until it isn't, and then the staff handles those moments so expertly, they can make the most chaotic turn seem somehow contained. The same applies to the happier turns—like a patient coming out of a coma after two months. It's a good day when a patient in the ICU makes progress, and today the nurses, attending physicians, and others offered their quiet high-fives, fist pumps, and hugs, sharing their own relief and joy. Meanwhile, at peace in the calm, quiet TICU cocoon, Tom got to sink quickly back to sleep, and we all got to rest easier that night, just knowing that he could. The cats cuddled close, and I surrendered to sweeter dreams than I'd imagined possible for a long time.

Tom: Interlude VII

There is a pecking order even in death. The blowflies know this, as do I. Of the necrophagous insects that feed on animal and human remains, the blowfly is among the first to arrive. Female blowflies laden with eggs are preferentially attracted by death volatiles, minute concentrations of gases like dimethyl trisulfide that are produced soon after death. Larvae hatching from their eggs emerge to start their feeding frenzy before the flesh fly lands. Ants or house flies typically appear in the next succession wave, paying their respects in a receiving line where the groom is devoured instead of the wedding cake. Sometimes, burying beetles accompany them, carrying tiny mites that snack on blowfly larvae. Since blowfly larvae produce ammonia that is toxic to the beetle, the mite keeps the larvae in check, an example of classic symbiosis. You scratch my antennae, and I scratch yours.

Blowflies prefer to slurp juices from bodies that are already decomposing, so the swarm hovers over me, waiting. Their beating wings create a deafening symphony; the sound of decay. A few of the anxious ones take turns sucking at my mucous membranes: eyes, nose, mouth, anus. My clothes are in tatters after so much time in the desert, offering me no protection. I keep walking, so they eventually depart en masse to search for a more accommodating victim.

I know a thing or two about flies. A different fly species infested me once. After spending three months in the Colombian jungle with several other graduate students in 1972, we returned home with more than postcards. Each of us was crawling with exotic parasites. The few mosquito bites on the back of my thigh had started out as minor irritations. Several weeks later they were infected masses the size of golf balls, then baseballs. I kept telling the doctor that whatever it was, it was eating me. I could feel it feeding, especially at night. Occasionally, it hit a nerve and my leg would jump like a marionette being manipulated by a sadistic puppeteer. At such times, I would slap at my bulging thigh and the thing would lie dormant for a little while. The doctor thought I was nuts at first, but when a one-inch pupae with three double rows of epidermal spines suddenly emerged while I was on the examining table, he was stunned. Turns out I had been afflicted with an infestation of Dermatobium hominis, *a botfly that ingeniously captures mosquitoes and lay its eggs on their underbelly. When the mosquito bites its host, the newly hatched botfly larvae crawl into the wound, feed on their host's flesh, and then pupate. Yum.*

The laws of nature are simple: eat or be eaten. Die and decompose. I made a conscious decision. I want to live.

I followed the hooded horde of swarming flies. Lead me to salvation or another carcass, but a place where life once flourished is one where there is at least a remnant of hope left. As I walk in perpetuity, I feel the sand of the desert becoming spongy beneath my toes. The air becomes moist; my nostrils flare. I swallow, my membranes hungrily absorbing every droplet. Up ahead is a bog; the sphagnum moss hangs over trees with skeletal steel limbs. I have been here before, but when? The light is dimmer now, the heat less oppressive. Above the swamp, a phosphorescent orb appears and beckons me. A will o' the wisp? The colors are violet, green, and blue, flecked with orange. I hear music and see its colors: I know this is synesthesia, which I haven't experienced in decades. Come closer, the orb seems to say. I draw nearer, and the drone of the flies becomes a hum.

Voices, not flies. People.

My eyes flutter, then open. My senses are suddenly bombarded with stimuli: bright light, a red plastic biohazard container, yellow gowns and blue gloves, antiseptic, the incessant beep of the cardiac monitor, the drip of the ventilator hose, the glint of the IV pole, the buzz and blink of the fluorescent light on the ceiling, the rasp of bedsheets on my skin, the brown mossy curtain of my TICU room. Bed 11. I stretch my spine. Although I feel my fingers, my feet are strangely numb.

I hear laughter, followed by a shriek of delight from a willowy figure that I instantly recognize is Carly. She rushes toward me, and I drink in her scent. I am instantly reminded of the day she was born.

I lift my head from what must be my pillow, and I reach out to her hand. I bring it to my lips and kiss it.

I am finally certain that I am alive.

24

SECOND-GUESSING

March 20–21, 2016

"What the fuck?!"

My mother would say she didn't raise me to have a mouth like a sewer. But when I stood at the foot of Bed 11 early in the morning of March 20, those were the first words out of my mouth.

The night before, Carly had been so thrilled after Tom had awakened and recognized her and Danny, however weakly. For the first time in months, we went to sleep excited and deliriously happy. And when I'd called the TICU this morning to check on Tom as I usually did every morning at five a.m., the charge nurse, Marie, said that the evening had been uneventful, in a good way. Convinced the nightmare was almost over, I practically skipped from the parking lot to the TICU, expecting to see Tom awake, at last.

What I was not expecting was what I saw. Tom was unconscious and his face was as pale as the white sheet that covered him. His skin was clammy and feverish. His heart rate was 135. Very tachy. And his blood pressure was dropping like a stone. I gowned up in seconds and rushed over to him.

"Tom! Honey! Can you hear me? It's Steff…if you can hear me, can you squeeze my hand or open your eyes?"

No response. All I could hear were the sudden alarms of the cardiac monitor. The numbers for heart rate and blood pressure were now flashing. Ray, his day nurse, instantly appeared from the neighboring room to see what was going on.

"What the hell happened, Ray?" I snapped. It wasn't yet seven thirty a.m., so his shift had started only moments ago. He was just as shocked as I. Ray ran a hand over his shaved head. The tiny bit of growth that was just beginning to show signs of gray against his dark skin gave him a fashionably grizzled look. By the look of how events were unfolding today, he was about to gain a few more gray hairs. My hair-trigger impatience didn't help, but his focus, as always, was laser-like on Tom. He hurriedly donned gown and gloves, adjusted Tom's pressor settings, and allowed me to look over his shoulder at the morning's lab values on the computer.

I yelped. "WBCs are 69,000! Is that a typo? Or a lab error?" By now I knew that the average range for an adult's white blood cell count was 4,500 to 11,000 cells per microliter. Tom's WBC count had increased slightly the day after phage therapy began, but that was expected. Today's WBC count represented an astronomical increase. Ray quickly scrolled through the numbers on the screen. The look on his face said it all.

"Holy crap," Ray said under his breath, giving his head a shake. "I've never seen a WBC count that high, and it increased from 14,000 to 69,000 just overnight. Let me page the critical care doc on call."

I felt numb. Inside, I was kicking myself. I should have driven back to the hospital last night as soon as I heard that Tom had woken up. As exhausted as I was, at least I would have seen him awake. If last night's brief glimpse of consciousness was the last Tom would ever experience, I had missed it. Dammit. Why had I been so naïve to think that Tom was out of danger, given how rocky his recovery had been so far?

I recalled the famous neurologist Dr. Oliver Sacks, whose memoir was made into the movie *Awakenings*. As a younger man, Dr. Sacks had been treating patients who had fallen ill during an epidemic of a neurological condition called encephalitis lethargica that struck an estimated half million people during the first quarter of the twentieth century. Its cause was a major medical mystery until recently, when studies implicated an enterovirus typically found in the gut, like polio. Patients with encephalitis lethargica typically developed Parkinson's-like symptoms and catatonia, a comalike state. Sacks discovered that by administering the neurotransmitter L-DOPA, he could miraculously awaken his patients from their fugue. Tragically, L-DOPA's effects were temporary. It was agonizing to watch these patients—along with their loved ones and their doctors who stood by helplessly—as they slipped back into a catatonic state, one by one. I knew what it felt like now to watch your loved one sink from sight that way. In Sacks's book, they saw it coming, and those last goodbyes were beyond gut-wrenching. At this moment, all I could think was that I might never have that chance.

While my brain tried to process the possible reasons for Tom's cataclysmic crash, I flitted around his room at lightning speed, trying to *do* something. I moistened the quick-cooling cloth and placed it on his forehead. I set the fan to high and positioned it on his face. But when I placed an ice pack under each of Tom's armpits, I noticed that several of the ostomy bags attached to his drains were unusually full. The bag draining the pseudocyst was collecting a dark brown murky liquid that was peppered with what looked like coffee grounds. I held up the bag to show Ray.

"The pseudocyst drain has been putting out one hundred milliliters of fluid a day, which has always been yellowish-brown," I nattered. When all else failed, I went into scientist mode; it gave me the illusion of control. My voice sounded mechanical, like Dr. Tempe Brennan in the TV show *Bones*. "If the bags were emptied on night shift like usual, then the pseudocyst has put out about five hundred milliliters of this dark brown gunk in the last eight hours. That's got to tell us something. And what are all of those brown flecks?"

"I'm no doctor, but those coffee-ground-like flecks are usually a sign of coagulated blood," replied Ray, as he inspected the contents of the ostomy bags and emptied them, saving a sample from each. Of course, I wasn't a doctor, either, but I suspected that the doctor on call or Chip would place an order that their contents be cultured. I took a photo of the ostomy bags with my cell phone and texted it to Chip, along with a shot of the cardiac monitor, which showed blood pressure 75/34, respiratory rate 35, and heart rate 121. Those hemodynamics were terrible. I tried not to jump to conclusions, but I couldn't help myself. The signs were all there. This was septic shock. Again.

But the startling sludge in the bulging ostomy bags reminded me of something Félix had described one hundred years ago. The intensity of the bacteria's reaction to phages "is of such violence that it must have been observed by many bacteriologists but had not been understood," he wrote. He'd referred directly to a lab in India, where a scientist had seen his bacterial cultures turn clear after twenty-four hours and referred to these as "suicide cultures." That, of course, depended on whose side you were on. One organism's suicide is another organism's victory party. Complete with exploding piñatas when the phages break through the bacterial cell wall and the virions come pouring out.

Chip texted back in record time, saying he would get dressed and head over to the TICU right away. His note said that he still thought the phages were doing their job. "I suspect this is something else," he texted. The question was, what? He arrived at the hospital within the hour, took one look at Tom, then looked at me. A sheen of perspiration glistened on his forehead. Other than this telltale sign, Chip appeared calm and collected. I later learned that he wrote Davey a note saying, "I think I may have killed Tom."

That made two of us.

Chip shifted into his own version of hyperfocus mode. He verified that the doctor on call had placed an order to change all of Tom's lines. This meant changing each of the thin catheter lines that had been inserted into Tom's blood vessels to deliver antibiotics or allow easy access for taking blood samples. Any could be the source of a new infection. Chip also confirmed that samples had been drawn from Tom's body fluids—blood, urine, sputum—as well as his drains, to culture any new organism that might explain his precipitous decline. Chip called the research pharmacy and asked them to send samples of the phage preparations to the micro lab to ensure that they had not become contaminated with bacteria from handling. Samples of feces and sludge from the drains were to be tested for blood. A blood sample was taken to test for cardiac enzymes to rule out a heart attack, since its early signs can include shock. The doctor on call had already ordered a CT to determine whether Tom had experienced a GI bleed—all *stat*. No time to spare. After all of the cultures had been obtained, Chip also increased Tom's traditional antibiotics. Patients in the ICU who are as sick as Tom have many places through which bacteria can invade, he explained, so they are always at risk for new infections that could be unrelated. When they take sudden turns for the worse, Chip said, it is prudent to look for new and different infections and to change antibiotics while waiting to see how the patient does clinically and while the cultures grow.

I paced the floor by Bed 11, which suddenly felt as oppressive as a prison cell. I could feel adrenaline coursing through my veins, but I had nowhere to direct it. I took a closer look at Tom's face. It looked like a death mask.

"I am really worried, Chip. Did you see that crap draining from the pseudocyst? It must be World War III in Tom's gut."

"I know. I am worried, too." Chip admitted ruefully. "The critical care team suspects that the phages are the culprit, but that's because they have no experience with it. I'm guessing that there's another microbial offender to blame here. We will get to the bottom of it, but in the meantime, are you okay if we put the phage therapy on hold for a day or two until we can determine the cause of the problem?"

"Yeah, I guess so," I replied, clenching and unclenching my blue-gloved hand. Maybe the endotoxin levels had been too high, after all, and this was where the whole thing would end. But what if it wasn't the phages or the endotoxin? Stopping the phage therapy was the logical thing to do under the circumstances, but I knew that this carried its own risk. Chip waited, seeing that I was torn.

"By reducing the selective pressure exerted by the phages," I finally asked him, "aren't we giving Tom's *Acinetobacter* more opportunity to develop resistance?"

Chip had already considered this. "That is definitely possible," he replied thoughtfully. "But we have the benefit of having two phage cocktails in our arsenal; that is a total of eight phages. No one really knows how much time we have, but I don't think the *Acinetobacter* has generated resistance to all of them within a few days. That said, the longer we wait to reinitiate the phage therapy, we do increase the risk that mutant bacteria will start to grow. That's exactly what happens when we stop antibiotics prematurely."

"But in the case of phages, things are a little different," he continued. "In theory, if they are behaving as we hope, they'll already have reached their *Acinetobacter* targets and will continue to replicate at the sites of infection—whether or not we continue to administer them through the drainage catheters and intravenously."

The hours ticked by while we waited for the battery of tests to be completed. His cardiac enzymes were normal, suggesting no evidence of an MI. But the CT revealed that Tom had had a recent GI bleed. That explained the coffee-ground appearance of the stuff oozing from his drains. Although another bout of septic shock had occurred, to find out what organism had caused it meant waiting for the report back on the cultures, which takes twenty-four to forty-eight hours. No quick mug shots.

Dr. Mims was on rounds again. He warned me that even if Tom awoke from his coma, I needed to prepare myself. He had been bedridden for so long that he could have permanent damage to his nerves or his muscles—neuropathy or myopathy—or both. How much more bad news could there be?

But there was more. A man Tom's size should contain nearly two gallons of blood, and the normal hemoglobin level should be at least 13.0. Each day in the TICU, his hemoglobin level was monitored to ensure that it hadn't fallen below the critical level, which was 7.0. When that happened, which it had on several previous occasions, there wouldn't be enough hemoglobin to carry oxygen to the body's vital organs, and a transfusion would be needed. Tom's hemoglobin was 5.5, meaning that he had lost almost half of the blood in his body. This was mostly due to his GI bleed, but for months now, his bone marrow had been busy making white blood cells instead of red ones. That also explained his zombie face. Dr. Mims had sent an urgent request to the Red Cross for three units of blood, but they hadn't arrived yet. Tom had received more than sixty units of blood since he'd fallen ill. Nearly eight gallons. His blood was becoming harder and harder for the Red Cross to match. I made a mental note to donate blood more regularly. One of the many reasons Tom was still alive was thanks to scores of anonymous blood donors.

The mood around the TICU had become somber, the hope for their miracle patient fading. No one would meet my gaze, which only made me feel worse. How was I going to tell Tom's girls that their dad had septic shock again? I was to blame because I had insisted on an unproven experimental treatment that was alive and multiplying inside him. My only consolation was that Tom still kept producing enough urine that the nephrologists were not pressing for dialysis. Just when I was feeling like I was coming apart at the seams, Davey appeared at the doorway to Bed 11.

"Hi, Sunshine," Davey said softly with a smile, and gave me a warm hug.

"Davey," I cried. "Am I glad to see you! What's your take on what is going on with Tom? How bad is this bleed, and what does it mean for his prognosis?"

Davey pulled up the CT scans on the computer and pointed at the black and white images on the screen. As he adjusted the view, a bolus of blobs morphed into various sizes before us, like time-lapse photography of a lava lamp.

"Chip and I went over the CT with the radiologist, so I will tell you what I know," Davey said, as he sat down backward on a swivel chair. "So, we are looking at a cross-section of Tom's abdomen. See that? That's his liver. And that small spongy glob? That's what is left of his pancreas.

He's lost about one third of it, but what's left does *not* look necrotic. Not dead—that's a plus."

"Okay." I nodded. "Those blobs could be Pluto and Jupiter as far as I am concerned. What's that cloudy stuff?" I pointed to an area around the pancreas.

"That's inflammation," Davey replied. "There's a lot of it, which is why he has so much ascites fluid in his abdomen. The fluid is putting a lot of pressure on his lungs, which is why he is still having problems breathing. Those cloudy regions around his lungs are pleural effusions. More fluid. But what I really wanted to show you is his gall bladder." Davey pointed on the screen to a melted hockey puck. Not what I'd imagined.

"Isn't the gall bladder supposed to look more like a sphere?"

"Yep," Davey replied, waiting for me to go on.

"So where did it go?" I asked, confused.

"Here," said Davey, holding up one of Tom's ostomy bags that was still collecting a steady stream of thick brown fluid peppered with coffee-ground flecks. "Some of it was probably excreted in his feces too, since it showed signs of old blood."

Yuck. I shuddered. I might never drink Peet's again. "So basically, the *Acinetobacter* infection eroded his gall bladder tissue so it just disintegrated and seeped out of him?"

"For lack of a better way to explain it, yeah," Davey grimaced. "But it's not the end of the world. You can live without a gall bladder. Lots of people do. Let's just see what grows from his blood culture and hope that it isn't the *Acinetobacter* again." What he didn't say was that if the superbug was isolated in his blood again, that would probably mean that the phage therapy wasn't working. And if Nature's ninjas weren't wielding their nunchucks by now, they never would.

But Mother Nature, Darwin, or both were on our side. By the next day, Tom's fever started to break and his WBC count decreased markedly. Chip dropped by Bed 11 in the early morning to check in.

"I thought I would find you here," he greeted me. "Looks like things are headed in the right direction again. We should get the results of the culture by this afternoon, but I am betting my white coat that the culprit isn't *Acinetobacter*."

Early that afternoon, Chip reappeared.

"Okay, let's have it," I demanded, trying to invoke some decorum or politeness, and failing.

Chip chuckled and his freckles stood out more than usual.

"The micro lab finished the cultures, and as I suspected, the phages are not implicated in Tom's latest episode of sepsis," he said. "His blood culture grew *Bacteroides thetaiodomicron*, a common gut bacteria, but a problem when it sneaks into the bloodstream like this."

Considering the possibilities, this was great news.

"Leave it to Tom to acquire yet another organism that no one can pronounce," I replied sarcastically. "This one sounds like it has been to a few frat parties. Now please tell me that this bug is susceptible to antibiotics."

Chip was equally gleeful. "Yup. It's sensitive to meropenem, and he's already on that, so we are covered," he responded. "That is likely why this latest episode of sepsis has already started to resolve. What's more, Monika, the fellow in Victor's lab, did some further sensitivity analysis on one of the isolates from Tom's drains that was taken right before the phage therapy. It was sensitive to minocycline. So, we are adding that to his regimen today. I have a feeling he is going to be back on track soon. With your permission, I would like to reinitiate phage therapy ASAP."

I agreed. We didn't want the *A. baumannii* to have an opportunity to start to grow again.

There was a driving-blind sensation and a hurry-up-and-wait rhythm to this whole thing that Chip seemed at ease with; me, not so much. There was no way in real time to monitor phage activity: Were they taking down huge satisfying swaths of *A. baumannii*? Making it to the obscure pockets of infection? Getting siphoned off and recycled by the liver? Languishing in some dead-end crevice where their bump-and-sniff style was stymied? You couldn't *know*. And the specimen they sent off every few days to the Navy lab took time to culture and analyze. Chip seemed to have a special gear—some sixth sense—he shifted into for complicated cases that called for him to work with the fits and starts inherent in data, the delays and the human factor. I marveled at this capacity.

"The reason I'm loath to stop now," Chip said, "is that we haven't yet measured the phage at the sites of infection and we don't know for certain that what we think *should* happen will *actually* happen in a patient like Tom."

His expression softened a bit, and he delivered his clinical decision with his down-home doc warmth.

"If a boat ain't rocking, don't sink it. Tom was clearly improving on phage therapy until twelve hours ago. Changing course suddenly by stopping phage therapy altogether is something that I want to avoid because his boat has not only not been rocking, it has been *rising*."

My bet was on Chip.

At the moment, he said, Tom needed to have a few of his drains upsized by interventional radiology. The glop oozing from his drains was so viscous, they were worried the tubes would get clogged, which could aggravate the infection. I left a voice mail for Carly, who was scheduled to replace Frances on the afternoon shift with her dad. I didn't want her to be blindsided when the phage therapy resumed. She texted me back a few minutes later:

I hope y'all know what you're doing. He's barely recovered from this last bout of septic shock.

She was right. Tom was not out of the woods yet, and there was no telling if plowing ahead again was going to be the biggest mistake of my life. Much later, I learned that this was the day when my nephrology colleague, Joe Ix, had asked his resident how Tom was doing. She told him, "Well, Chip thinks he's getting better, but nobody else does."

25

NO MUD, NO LOTUS

March 22–31, 2016

The next morning, I lay in bed and hit speed dial an hour earlier than usual, at four a.m. Despite the pitch-black, I could feel the button as if it were Braille. By chance, TICU was lucky number 7. I would take whatever luck came my way these days.

"Thornton ICU..." a female voice answered, which I suspected was Marie, the charge nurse on night shift.

"Hi, I am calling to see how my husband, Tom Patterson, slept through the night." I continued to lie in bed in the dark, stroking Paradita's furry stomach with my free hand. The whole herd was with me: Bonita was perched on my hip, and Newt was curled behind my legs, lightly snoring. At least one of us was getting a good night's sleep. I knew that with these four-legged alarms, there was no chance of me getting any more *Z*s.

"It's still happening," said Marie. I could hear her typing on a computer keyboard while she spoke.

"What is?!!" I exclaimed, jumping out of bed. Upended, Bonita, Paradita, and Newt scattered in all directions. If Tom's interventional radiology procedure had lasted all night, something had gone terribly wrong. Why hadn't they called me?

"The *night*," she replied calmly. "The *night* is still happening. It's four a.m."

"Oh, yeah," I replied, trying to dial down the edge in my voice. It *was* a time when most people were sleeping. "So is he?"

"Is he what?" Marie asked benignly. Was she joking? I couldn't tell.

"Sleeping through the night!!" I snapped. It felt like we were caught in the *Twilight Zone* version of "Who's on First?"

"Like a baby," Marie replied. "Now go back to sleep—and that's an order!"

A few hours later, I shook off my snarly attitude and drove to the hospital. The earlier call had left my nerves jangly, which seemed irrational, but I was still reacting to the day before, when I'd assumed Tom was fine, based on all reports, and arrived to find all hell had broken loose. Now, although I wanted to hear good news, it was hard to trust it. The radio was playing Louis Armstrong's "It's a Wonderful World," one of my old-time faves. I sang along, reflecting that Tom would have been six feet under by now if it weren't for the phage community and all the doctors and nurses who worked so hard every day to keep him alive. People like Marie, who'd kept her calm when I'd snapped at her. It *was* a wonderful world, or had been anyway. Maybe it would be again. If only Tom could pull through.

Through the atrium, past the line of towering palms standing at attention, and with just enough time to say a little prayer in the elevator—*Please, God, give me the strength to get through whatever this day brings me*—the doors opened on the new day, a new shift in the TICU. And day 115 since Tom had gotten sick in Egypt. I checked the whiteboard behind the nurses' station and saw that Chris was assigned as Tom's sole nurse today. It had been a week since he had stood by Tom's side as the phage therapy had begun with the Texas cocktail. I gowned up and stepped in, and Chris greeted me with a hug.

"I just finished catching up on Tom's chart. I know you've gone through hell these last few days," he said kindly. "But things are already looking better today."

Tom's blood pressure was stable, so they were able to lower the pressors, and his heart rate was down below 100 for the first time in days, Chris said. "And," he added, "he seems to be stirring."

I almost shrieked but managed not to.

"Are you serious, Chris?! No—*really*? Are you *sure*?" I said. "Tell me you're not shitting me."

Chris smiled. "I shit you not. Watch this." Chris approached Tom's figure on the bed, and leaned down toward his ear. "Tom, it's Chris again. I just need to ask you a few more questions. Can you use your left hand to squeeze mine?" A few seconds passed, then I saw Tom's hand move almost imperceptibly and give Chris's a faint squeeze. I gasped.

I breathed into Tom's face. "Honey, honey, it's me. It's Steff. Can you open your eyes? I have missed you so much!" *Please, God, please let him wake up.*

I waited for what seemed like an eternity. Tom's eyelids fluttered, but they were glued shut with yellow crust. Chris wiped his eyes with a warm cloth. I stroked Tom's cheek with my blue-gloved fingers, willing his eyes to open. Then suddenly, they *did*. His eyes were slightly unfocused at first, but when he looked at me and smiled faintly, my insides went to jelly. Tom's lips were gooey and cracked and his teeth were furry, but I couldn't care less. My husband was back. As I leaned over him to give him a little kiss, his lips puckered to kiss me back and he nuzzled my neck. I heard a whimper and realized it was coming from my throat. Tears streamed down my face. Tom looked around the room, bewildered, then he looked back at me. I could tell that he didn't know where he was and that he was surprised to see me crying. But at least he knew *who* I was. Or at least I thought so.

"Welcome back, babe," I whispered, wiping away my tears. "You are in the Thornton ICU. Do you remember getting sick in Egypt?"

Tom stared at me blankly and shook his head no.

Oh no. Here we go again. How many times had we gone over this? Was he going to have amnesia? Or worse, permanent brain damage? Or was this just another round of the old ICU psychosis? I looked at Chris with alarm.

Chris read my mind and put his hand on my shoulder. "Don't worry," he said quietly. "It's common for patients coming out of a coma to be disoriented or to have no memories of how they got here at first. Just be patient. For now, just tell him the basics."

That made sense. We couldn't overwhelm him. Deep, cleansing breaths.

"Honey, you went into septic shock after your drain slipped inside your gut and the superbug infection spread throughout your body." So much for the basics. I took another deep breath and began again, my voice trembling. "You were in a coma for...for a while." Yeah, as in off and on for nearly two months, I thought. But maybe he didn't need to hear that right now.

"Chip and I gave you the experimental treatment I told you about before. Phage therapy. After a bad turn, we think it's working. Things were touch and go for a while, but you are going to make it now, babe! I love you so, so much. Just please, hang in there. You're doing great."

Tom still looked confused, but my explanation seemed to placate him. He squeezed my hand and his lips moved to speak, but he couldn't talk since he was still on the vent. His hands thrust almost reflexively to his neck, to pull at the ventilator hose. But after so many days of the same routine, Chris and I were ready. I grabbed one of Tom's hands and Chris grabbed the other before he could yank the ventilator tubing out of his neck. Tom glared at us, me especially, his eyes full of betrayal.

Chris gently took Tom's hand in his and crouched down so he could look Tom in the eye. "Tom, buddy. You had a tracheostomy so we could put you on a ventilator. To help you breathe. Just trust me, OK? Keep your hands away from your neck and try to relax."

Hearing this, Tom stopped straining and let his head fall back on the pillow in frustration. His lips started to move again. As he processed Chris's words, he gurgled, suddenly remembering all over again that he couldn't talk. I waited to see if Tom could mouth the words to whatever he wanted to say. Maybe something profound or momentous, suddenly freed as he was from the months-long prison of the coma.

Water, Tom's lips said. Not profound. Obvious.

I was almost positive he wasn't allowed to drink anything, but after the death stare Tom had just given me, I wasn't going to be the bad guy. That was Chris's job. I asked Chris innocently, "Can he have water?"

Chris shook his head no, vigorously. "Not while he is on the trach," he replied, giving me a sharp look like I should know better. "But he can have this." He approached Tom with one of the tiny sponge lollypops we could dampen and use to wet his mouth. Tom's face blanched and he turned his face away. Selective memory, I guess.

I heard a voice from the doorway. It was Dr. Kim Kerr, the director of the ICU. "And no, he can't have ice chips either," she said, wagging her finger at us. "But the fact that we are even discussing this has just made my day!" She was beaming. Dr. Kerr had just been named Physician of the Year, and I knew why. She made every patient and their family feel special. Behind the reception desk, Joe, Meghan, and Ray had heard the news. They were glowing, and gave me a wave and thumbs-up before going back to their duties.

Marilyn joined Dr. Kerr at the doorframe to Bed 11. "You saved his life, you know," Marilyn said softly.

"We all did," I responded, my voice catching. "We all did. Your team here is awesome. We can't thank you enough." I stepped back to take a photo of Tom with my cell phone. I wanted to remember this moment forever.

Behind her was Joe Ix. "Well, I'll be damned. Look who's awake! My God." Joe's jaw practically hit the floor. "It's a miracle. Really, I have never seen a turnaround like this before. And phage therapy. Just...wow. Congratulations, Steff! At this rate, he won't need dialysis. My team will keep monitoring him for the next few days, just in case." And with a little wave, he joined the nephrology fellows and proceeded on his rounds.

Within the hour, Frances arrived. Her face was pale and drawn and I suddenly noticed how thin she had become. She came to a dead stop down the hallway as she approached Bed 11 and looked at me in a panic. She had immediately drawn the same wrong conclusion I had earlier, assuming the worst when she saw the throng of doctors and nurses clustered at the doorway of Bed 11 to bask in the glow of our medical miracle. In an ICU, more often than not, a crowd like this usually meant code blue, or that a patient had died.

"Fran, your dad just woke up!" I exclaimed as I pranced around the bed, ending up in front of her. We gave each other a bear hug. I could feel her rib cage, and I knew her heart was beating wildly within it. "He's sleeping now, but it looks like he is finally on the mend."

"Do the doctors say he's going to live?" Frances asked me in a quiet voice, in case her dad could hear.

"They haven't said that exactly, but there is more hope around here than we've had in a long while," I replied carefully.

Frances sat down silently, taking it all in. After a few minutes, she picked up a small whiteboard that was used to communicate with patients who couldn't talk. With a black marker, she started drawing a picture of a flower. I looked over her shoulder as she wrote above it:

No Mud, No Lotus.

She looked at my puzzled expression. "It's one of my favorite parables by the Vietnamese philosopher Thich Nhat Hanh," she said, as she wrinkled her nose. "I'm sure I'm mispronouncing his name, but basically he says that just like the lotus can't bloom without mud, we can only achieve enlightenment if we embrace suffering, rather than trying to avoid it."

I looked at her in wonder. Like Tom, Fran and Carly constantly surprised me with a kind of grounded energy at times when it would be easy to be overwhelmed by the emotional charge of the moment. At least it was a positive charge, for the first time in a long time.

"We should all be well on the way to nirvana by now then!" I replied.

Tom reopened his eyes, as if on cue. He and Frances exchanged a tender embrace, and there were more hugs and tears all around. Minutes later, Chip appeared, wearing his white coat over a pair of Dockers and a plaid shirt. "I ran into Kim in the parking lot," he grinned, taking in the scene.

"Honey, Chip led the protocol for the phage therapy that saved your life," I told Tom proudly. "What do you think of that?"

Tom looked up, and gave Chip two enthusiastic thumbs-up and a big smile.

Chip clapped his hands together and let out a belly laugh. He opened his mouth to say something, but had to pause as his eyeglasses steamed up and he removed them to give them a wipe. Gathering himself, he finally responded to us.

"Moments like this remind me why I became a doctor," he said, his voice wavering. "I couldn't be happier!"

••••••••

Over the next week, Tom's clinical status continued to improve and life in the TICU returned to its own kind of normal. One morning, Carly took the early morning shift so I could sleep in. When I arrived at the hospital later that morning, the commotion from Tom's room as I approached made my heart skip a beat until I heard the excitement in the voice of Will, the respiratory therapist.

"Oh my God! He's locked in! Is he in the tube? I can't see!" It was Will, shouting from inside Tom's room. Excited—in a good way. In their shared surfing world, the "tube" is the barrel in a breaking wave, where you want to be.

"Hey! I could hear you guys hooting all the way through the glass walls, for cryin' out loud," I said, gowning up and stepping in. Will was on tippy-toes trying to get a better view. He absently fiddled with the settings on Tom's ventilator, but he was riveted to the TV screen overhead. So were Carly and Tom.

The three of them were glued to the Rip Curl Pro Bells Beach surfing competition. White-knuckled, Tom gripped the sides of the bed in anticipation as they watched an Aussie, Matty "Wilko" Wilkinson, attempt to clinch first place. Carly craned her neck to see as she rubbed cream on Tom's feet, which continued to slough off papery snakeskin. Wilko was cutting back, expertly carving out each section of the wave. It would be a long time before Tom would be back in the water, but Carly had taken up bodysurfing and was keen to take him out to catch a few sets. The mutual father-daughter delight might as well be genetic; in SoCal, surfing isn't just a sport, it's a way of life.

"Hoooo!" Will bounced up and down on his cross-trainers, exuberant. He held his hands out and waved his arms gracefully, positioning his feet as if he were riding a wave of epic proportion. "Will you look at that! I can't remember the last time a goofyfoot took a title. And he's been chucking his tail all week. Dude, what a ride!" Will put his right foot forward ahead of his left and teetered on his imaginary board, mimicking a surfer who sported a "goofyfoot."

"Will, speak English, will you?" I teased, leaning over to inspect the ventilator settings. Then I promptly had to shift to medicalese, everyone's second language here. "Hey, how long has Tom been sprinting off the vent?"

Dr. Kerr had explained to us that the lungs are a muscle and Tom needed to do breathing exercises every day to give them a workout. They call it sprinting. As much as people had told us this would be "a marathon and not a sprint," Tom was sprinting after all. Or at least his lungs were.

Tom held up two fingers.

"Peace, baby?" I grinned, and ruffled what was left of his hair.

Carly chimed in. "Nah, he means, two *hours*. He's been rockin' it this morning."

••••••

Breathe. In through the nose. Out through the mouth. In, out. In, out. Relax.

Each day, Tom breathed off the vent longer and longer. One morning after rounds, Dr. Fernandes stopped by. "Today is your lucky day," he said, snapping on his gloves. "We are going to extubate." Tom stared at him, uncomprehendingly.

"Hon, he's going remove the respirator tubing from the tracheostomy hole in your throat. Finally!"

"Yes, and then we cap the stoma, just leaving the cuff," said Dr. Fernandes. He turned to a resident, Christine, who prepared his equipment.

"Leave it in—why?" I asked.

"We don't want the stoma to close up quite yet. Just in case we need to put him back on the respirator later," Christine explained.

Tom's eyes widened and he shook his head wildly. Vent or no vent, I knew he was thinking: *Not bloody likely. I'll tear the damn thing out myself!*

Dr. Fernandes could tell this wasn't exactly the news we were hoping for. "We don't expect any more problems, but given the course of this illness and the—er, unconventional aspects of this phage treatment, we need to be very cautious. When it's time, we will remove the cap and the trach hole will close up very quickly," he reassured us. "It's almost like having your ear pierced."

Yeah, not quite. I reached for my left earlobe, pulling it absentmindedly.

Within a matter of minutes, Tom was no longer technically on life support. He was breathing fresh air. Well, almost fresh. The smell of urine, antiseptics, and other hospital accoutrements wasn't exactly aromatherapy. And Tom hadn't had a real bath in four months.

Will had been off for a few days and was in awe to see that Tom had been taken off the vent. He put his hand on his own collar, cleared his throat, and preened, as if he were wearing an ascot like Tom's stylish stoma cuff.

"Dude, you look awesome!" Will gushed. "You're going to be back in the tube in no time!"

Coming from a fellow surfer, that must have been music to Tom's ears. He gave Tom the hang loose sign with his thumb and pinkie, and disappeared into Bed 12.

After hearing that Tom had been extubated, Chip texted congrats on the good news. His message included a smiley face emoticon. His sense of humor was recovering, too. I imagined Connie had probably showed him how to use emojis so he could impress his grandkids. Signs of life all around.

Mail call now included get-well cards I could finally read with Tom, and one day it brought a small purple stuffed toy. Tom looked ready for a nap, but he gave the toy a quizzical look—it looked a lot like a spider.

"Babe," I said, "it's a phage. A toy phage, from the team at AmpliPhi that donated one of your phages!"

He nodded, amused, and then closed his eyes gently and dropped off to sleep.

No rest for the weary those next few days. Tom had work to do. He was learning—relearning—to swallow, drink, and talk. A speech pathologist came in to work with him twice a day. He had to practice moving his cheek muscles, which had atrophied because he hadn't used them in so long. Before he could be given water or ice chips, he had to pass the three-part "swallow test," which went like this: The first day, they give you an ice chip. The second day, they give you a spoonful of applesauce. Finally, on the third day, they give you a cracker. Don't gag. Swallow. He swallowed and got an A+. Now he could graduate to a liquid diet.

Talking doesn't come easy after a tracheostomy, either. He had to learn to speak with a special valve, which attaches to the outside opening of the tracheostomy tube and allows air to pass in the stoma, but not out. The speech pathologist had him practice vowel sounds, first, then consonants. After a few days, he was ready to say his first word.

"Steff," he whispered.

I had been sitting beside to cheer him on in these lesson times, but the first time I heard Tom say my name again, I bawled like a baby. Tom patted my hand and we laughed; for the last several months it had always been the other way around. It felt so good to laugh again, even if Tom's came out like a wheeze. Managing a sentence was much more difficult. Those sounded more like a croak.

"It's okay," I told him. "I love you all the same, my froggy prince."

"Knock, knock," said a familiar singsong voice as the door to Bed 11 slid open. It was Amy, the physical therapist. Today, her long hair was in a ponytail and she wore blue scrubs beneath her yellow gown. "Is now a bad time?"

Tom grimaced. Now was always a bad time for physical therapy. All work, no play. But he knew it was the only way to recover enough to be discharged.

Amy stood at the foot of his bed. "So, what do you feel like doing today?" By now, she knew that giving Tom this modicum of control was the best way of motivating him.

"A walk?" Tom croaked.

Amy gawked. "Walk? Are you pulling my leg?"

"Walked... to the end... of the hall yesterday," he croaked back haltingly. And grumpily.

Amy looked at me in dismay. I looked at Tom. The only place he'd walked since Thanksgiving was in his own mind—what I later learned was the trail to his family cabin, the desert, the swamp, and maybe joining Carly on her journeying visions of their walks together.

Amy sat down on the corner of his bed and pulled her ponytail tighter before looking at him. "Tom," she said softly. "You haven't walked in four months."

Tom looked at me in alarm. All I could do was nod. She was right. He was... mistaken.

God knows what he was thinking, but he wasn't thinking straight yet. That much was clear.

Tom: Interlude VIII

It's not like the movies. In the Hollywood version, the patient awakes from a coma, sits up in bed, stretches their arms like they've had the nap of the century, and asks what's for dinner. What you may not know—if you're unlucky enough to fall into a coma yet lucky enough to wake up from one—is that you may have to be taught how to swallow again before you can eat, drink, or talk. Your brain is kind of fuzzy. Your muscles forget how to do things. You might not even know where your hands and feet are, or how to move them. And at first, staying awake even for a few minutes is an effort.

At least that's the way it was for me.

Waking up was gradual. In the beginning, everything was a haze, like I was a mummy being unwrapped by an unseen hand, layer by layer. If someone was standing right in front of my peephole, they were a part of my world. If they were offstage even a tiny bit, they did not exist. But as I gained greater awareness of my surroundings, there were times when my body was like a tree struck by lightning. A simple touch from Steff or resurfacing of a childhood memory felt like an electric shock, setting off a cascade of neurons that would light me up, branch by branch and leaf by leaf. It was jarring, but I knew it meant I was alive, and hopefully that I was getting better. I could tell by the endless stream of visitors, family, friends, students, doctors, and nurses, many of whom who were crying, that they had not expected me to live.

How long had I been out for the count? I tried to read the date that the nurse, Chris, wrote on the whiteboard on the wall, but it was too blurry. Surely it had only been a few days? I wasn't sure. There were dozens of get-well cards taped to the walls, along with a bunch of sagging birthday balloons that were clearly past their prime. Whose birthday was it? Mine wasn't until February. And there was a contraption on my neck Chris said was a ventilator. Itched like hell.

Christ. I ran my hand across my face; my beard was long and ragged, like Rumpelstiltskin. My hair hadn't been this long since the sixties, and I could feel hollows in my cheeks and beside my eyes that weren't there before. But my arms, legs, and belly were puffy. From the neck down I could pass for the Pillsbury Doughboy. Davey explained that I had edema, which meant my blood vessels were leaking fluid into my tissues. The docs were giving me a diuretic to rid me of the extra fluid, and I had a catheter in so that I didn't have to pee in a urinal every few minutes. I was falling apart between the seams.

The insides of my mouth felt like the Sahara. Maybe because I had just been walking in the desert for one hundred years. Or had I? I knew no one could survive in the desert that long.

What was real, and what was I imagining?

Before, I could have sworn there were hieroglyphics on the walls, but that was in my head. But that toy stuffed phage on the windowsill? Call me crazy, but it looks an awful lot like an Ankh, the Egyptian symbol for Life.

26

THE DARWINIAN DANCE AND THE RED QUEEN'S PURSUIT

April 1–6, 2016

I wish my five a.m. call to the nurses' station had been an April Fool's joke, but it wasn't. I had lain in bed in the dark using my cell phone as a night light, and hit 7 on the speed dial. Tom's night nurse, Larry, was in his room when I called, so I held for a few minutes listening to Muzak before ending up talking to the charge nurse, Marie.

"We were debating earlier about whether to call you or not," Marie admitted. "Tom had a rough night. He desatted at around midnight. He's back on full pressure support."

"What?!" I spluttered. *Desat* meant the oxygen levels in his blood were too low—again. We had started to see signs of that yesterday during PT. I pulled out my ponytail band and gave my head a shake, trying to wake up. "Does he have a fever? Do you have a WBC from this morning's chem panel? Is he back on the vent?"

"Yes, no, and no," replied Marie, handily. His temperature had spiked at 103.5. She didn't have the full chem panel, but his WBC count had been trending up through the night. I got the picture. Septic shock. Again. What was this now? The sixth time? I'd lost count.

Fuck, fuck, fuck.

Carly was still asleep. Danny and Frances had returned to the Bay

Area the day before. We'd thought Tom was out of the woods, at least enough for the kids to begin to resume some semblance of their own lives again. They'd each left a lot hanging to be down here. I looked at the mountain of dirty clothes that was overflowing from the hamper and made a split-second decision what to wear. Stooping down to pull out my favorite gray hoodie, I gave it a sniff; it could live to see another day. I threw it on with a pair of velveteen leggings and padded to the kitchen. Within five minutes, I nuked yesterday's leftover coffee and poured it into a travel mug, fed the kitties, and grabbed my car keys. Breakfast could wait. Besides, my stomach was doing flip-flops.

When I entered the TICU, the whiteboard behind the nurses' station showed that Joe was assigned to us again today. He was 1:1; one nurse for one patient. That assignment was reserved for the sickest of the TICU patients, which Tom hadn't been for weeks. As I approached Bed 11, my worst nightmare unfolded.

The room was empty. No Tom, no bed. Rosie was in there mopping the floor. There were only two possibilities. Tom was dead, or he was getting a procedure of some kind. If it was the latter, it was serious enough that they couldn't do it bedside.

"Rosie, where's Tom?!" I asked her in a panic. She looked at me and shook her head.

"Don't know," she replied sadly, looking at the floor, and kept mopping.

I ran back to the nurses' station. It was one of the few times when no one was there. I looked at the clock.

Shift change. *Dammit.*

I paced the floor for a few minutes, and then pulled out my phone to text Chip.

Before I could hit the Send button, Joe entered the TICU and saw me standing there stricken.

"Joe! Where's Tom? What's happening? Is he...??"

Joe grabbed both my hands in his and looked me in the eye.

"Steff. Calm down," he said quietly. "He's getting a CT."

Oh. That made sense. If he was septic, we needed to figure out the reason, and quick. But another CT meant that they had likely given him contrast, the imaging dye that was used so that the radiologists could read the films better. Contrast was hard on the kidneys, and that was the last thing Tom's kidneys needed right now.

My phone buzzed. It was Chip texting. *Are you in the TICU? I am coming up.* No smiley face emoji this time.

Chip strode through the double doors of the TICU a few minutes later, and we conferred in Tom's empty room.

"I just read the CT with radiology," he told me in his matter-of-fact tone, the same way he would tell me that he had eaten scrambled eggs for breakfast. "Tom's biliary bed drain has migrated into the hepatic parenchyma."

I blinked. I only understood two words of what he had just said, but it was enough. *Drain migrated.*

"You mean another one of his drains slipped?" I asked him.

"Yes," Chip replied. "Drain three." Of the five. "IR is going to reposition it, *stat*."

"Oh, for crying out loud. So, what now? After drain one slipped, he went into a coma and we almost lost him," I reminded him. As if he could forget.

"Yes, but that was before the phages," Chip said. "He's stronger now. Although I have to talk to you about that." His expression darkened. "Theron and Biswajit ran some studies on Tom's more recent *Acinetobacter* cultures. Theron called last night to tell me that his bacteria is now entirely resistant to the Texas phage cocktail, plus it's also resistant to every phage in the Navy's cocktail except one. As of today, we will stop using the Texas cocktail and start administering the existing Navy phage into Tom's abdominal drains, too, rather than IV alone."

Within a few weeks, almost every phage in the two cocktails, and each cocktail as a whole, had lost activity against Tom's isolate. No more shiny halos.

"Oh my God," I whispered. "How could it become resistant so quickly?" But Scientist Me knew why. We think of evolution over eons, but in the microbial world, it can happen overnight.

"It's the perpetual Darwinian dance," said Chip. "The pressure from the phages selected for mutations in the bacteria that could evade them. Given *Acinetobacter*'s doubling rate, their microbial army had plenty of time to come up with a new escape mechanism. I have a hunch that what they've done is drop their capsule. And if they have, they might have left themselves open to a new attack."

Chip was referring to the protective sheath that some bacteria, including *A. baumannii*, have covering their cell wall. Capsules can be a virulence factor because they give the bacteria an edge with genetic tools to change or block receptors, strengthen the cell wall, or tweak other features to enhance its prowess and defend against intruders, like antibiotics and phages. The capsule also carries water to keep the bacteria hydrated and helps pick up new resistance genes from other organisms the bacteria encounters. Dropping its capsule could be a big deal.

But I was only half listening. I was now officially freaking out. Even if it had dropped its capsule, the *Acinetobacter* had won this round, fully resisting all but one phage. And without those phages to keep the pressure on, the *Acinetobacter* could surge again.

Since Tom was already having another septic shock episode, maybe the superbug had already surged again. It might be too late for a new round.

"Isn't there anything else we can do?" I asked Chip, pleadingly.

"Already taken care of," Chip replied in his clipped tone. "Theron got the green light from his CO to start another phage hunt, this time using environmental samples."

I drew a blank. Beside me at the nurses' station, Joe looked up from his paperwork.

"He means his commanding officer," he replied.

I nodded, remembering that Joe had been a military nurse.

"So it's full steam ahead?"

"Yep. All hands on deck—the poop deck," Chip kidded in an exaggerated version of his southern drawl. Quintessential Chip. But in this case, miraculously, the bureaucracy was working in our favor.

Chip resumed a more serious tone. "The Navy's last cocktail was drawn from their phage library. Now Biswajit has already started screening phages obtained from local sewage samples. If he finds any, he'll need the time to characterize them and purify them to the FDA standards. Then we get a new eIND for the new cocktail, and we're good to go. We could have a new phage cocktail within a week."

A week. Tom could be dead by then. I didn't want to seem ungrateful, but in the ICU, battles for life were lost in the space of a heartbeat. A week felt like forever. My face fell, belying my terror.

"If I'm right, the timing should work out okay," Chip reasoned, trying again to assuage my fears. "What I was getting at a minute ago is that the more recent *Acinetobacter* cultures from Tom's drains look different when they are cultured, which is why I suspect they've dropped their capsule to evade the phages. Biswajit and several of his colleagues from the Navy and Army just presented findings on a similar arms race between *A. baumannii* and *A. baumannii*–specific phages. The *A. baumannii* capsule contained the phage's receptor for entering the cell," Chip paused to rotate his thumb and forefinger clockwise, like he was turning a key into a lock. "So, once the bacteria mutated and lost their capsule, the phages couldn't enter the bacterial cell anymore. But the bacteria suffered a genetic penalty."

I finished his sentence. "The *A. baumannii* mutant strain was less pathogenic now. The Red Queen hypothesis, right?"

Chip raised his eyebrows. I gave him a smug little smile; I had read the conference proceedings, too.

"Exactly," he said. The Red Queen hypothesis refers to *Alice in Wonderland* and the imperious Red Queen character, who tells Alice that "it takes all the running you can do, to keep in the same place." Evolutionary biologists used this literary analogy to explain how all creatures—predator and prey—must constantly adapt and evolve simply to survive.

With his background in evolutionary biology, Tom would love to hear about this, if it wasn't for the fact that his body was the military theater for the invisible battle playing out.

Just then, the double doors to the TICU swung open and a transport team pushed Tom's bed through on their way back into Bed 11. Tom's face was flushed and feverish, and he looked exhausted. When he saw us, he smiled weakly. As Joe hooked him back up to the cardiac monitor and adjusted his pressor settings, Chip and I took positions on either side of his bed.

"Oh, honey," I murmured and squeezed his hand. "I'm so sorry you are going through this again."

"I am, too, Tom," said Chip. "But we already know what's wrong. The drain in your gallbladder took a little walk through your liver. IR is going to reposition it for you today, and you should be back up and at 'em in no time."

Tom looked at Chip and his lower lip quivered.

"I'm not sure I can take much more of this," he whispered laconically, squeezing my hand tightly. I knew he was on the verge of tears.

"I know," Chip said quietly, his voice full of sympathy. "You have endured one of the most harrowing clinical courses I've seen in all my thirty years as a physician. It's been your strength and spirit, not just the phages, that has kept you alive." He put his hand firmly on Tom's bony shoulder. "You will get through this."

Tom nodded almost imperceptibly. "Thank you," he whispered weakly. "I really needed to hear that."

Chip and I exchanged a look. Both of us knew that this was not the time to tell Tom that his infection had developed resistance to the phages he was getting, and that the *Acinetobacter* was morphing right now, right there inside him. We could only hope that if the newer mutant bacterial strain proved to be less pathogenic, his immune system might rally and keep it in check while the Navy team tried to generate a new phage cocktail.

That afternoon, Dr. Picel and his team in interventional radiology readjusted Tom's drain. During the procedure, I imagined him threading the tube carefully through Tom's liver to what was left of his gallbladder, where it could hopefully continue to drain off biliary sludge. I wondered if, on days like this, Dr. Picel felt more like a mechanic than a doctor.

Tom pulled through like a champ, but he continued to spike a fever for the next few days. Forty-eight hours later, his blood culture came back as expected: the dreaded *A. baumannii* had refused to surrender.

27

THE LAST DANCE

April 7–May 31, 2016

Biswajit's new phage didn't come from the Navy library or some exotic port at sea, but from the murky waters of a sewage treatment plant in Laurel County, Maryland. For all the sophisticated science in play back at the lab, this step in the phage harvest was more Huck Finn than high-tech: an empty half-gallon plastic milk jug filled with rocks and tied to a rope. Biswajit and Matt, a lab technician, trekked out to the treatment plant, one of their favorite hunting grounds. Perched at the edge of the water, Matt held the end of the rope and flung the jug out, let it sink and fill with the scummy water, then hauled it back and emptied the brackish brew into a half-dozen large, capped test tubes. If not exactly panning for gold, then fishing for phage.

And they found a good one. It was officially designated AbTP3Φ1, but quickly earned the nickname Super Killer—the perfect match against a superbug.

In the lab, when Super Killer was dropped into cultures of Tom's bacterial isolate, the clear plaques that blossomed were the answer to everyone's prayers. I remembered the German medics, Anneke and Inge, who'd come to Luxor to fly Tom out, and how, in his delusional state, he'd seen them as angels in combat boots coming to the rescue. They'd earned their wings. Now maybe this new phage could earn its halos.

It wasn't fancy. In contrast to the phages in the original cocktails, which were myophages and had the longer, iconic contractile tails, this was a small, squat podophage with a short, noncontractile tail, consistent with the ancient Podoviridae family, subfamily Autographivirinae. So much for evolutionary upgrades. It added only one more phage to the cocktail mix, but initial tests for its activity against *Acinetobacter* indicated it was an efficient little bugger. Super Killer AbTP3Φ1 effectively killed the original bacterial isolate (TP1) and the two mutant strains that had followed (TP2 and TP3), which had become resistant to the phage in the first-round cocktail. Biswajit suspected that it targeted a different receptor than the myophages, which would make the new cocktail highly potent against *A. baumannii*. Not only that, the new phage also proved to be synergistic with one of those first-round phages, reviving its lagging activity against Tom's isolate. Together they stopped the *A. baumannii* in its tracks, at least in the lab—the Biolog assay panel charting the growth curves of each phage and the isolate showed a flat line for the bacterial growth.

Once again, after this second-generation phage cocktail was cleaned and prepped and administered—about ten days later—we watched and waited, holding our breath that Tom wouldn't have another relapse. Hours passed, then days. Tom developed sepsis again, but this time he rebounded quickly. His immune system was getting stronger, and he no longer needed blood transfusions. Other good news began to emerge in the weeks ahead. The culture from his pseudocyst and other drains revealed lighter bacterial growth of *A. baumannii*. Other tests also indicated that the *Acinetobacter* present now in Tom's isolate was a shadow of its former self. Chip had been right.

I hoped the new phage would be enough to give Tom's *Acinetobacter* a one-two punch. As we would eventually discover, Super Killer delivered that and more.

****

Every day in the TICU was the same old routine. Vitals. Meds. Blood draw. Rounds. PT. Bed bath. Sleep. Phages…

Of course, he was the only patient to get those.

"Billions of phages!" Tom crowed. "Talk about going viral!"

We all groaned. Tom's sense of humor had been one of the first things to recover.

None of us cared how weird this treatment was, as long as it worked. And it was working.

Chip, the Navy and CPT teams, and others exchanged constant reports from the battlefront that Tom and I could only imagine. They were monitoring activity underway in the trenches—inside Tom, that is—through the continuing data from blood draws and isolate samples. The Navy's Biolog was dutifully cranking out graphs and images every fifteen minutes as each round of Tom's isolate and the new phage cocktail were analyzed for the bacteria's sensitivity and the phages' virulence. The Darwinian dance was in full swing. Round by round, with the Super Killer in the mix now, the cocktails kept the selective pressure on, and the *Acinetobacter* was losing its grip.

By mid-April, Tom had improved enough that he no longer needed pressors to keep his heart pumping, and the TICU docs downgraded him to "intermediate" care status, meaning that he no longer needed "intensive" care. Never had a downgrade been so welcome. Given his track record in and out of medical mayhem, however, they opted to keep him in the TICU where the nurses could keep an eye on him a little longer.

We weren't the only ones breathing a sigh of relief. I'd stayed in close touch with Ry, Jason, and the CPT team. Jason had confessed at one point that even if the phages fully cleared the bacteria, he worried that Tom's body would never be able to recover. They'd been holding their collective breath, worried all along that we could still lose Tom. They'd just focused single-mindedly on keeping the hospital supplied with the Texas phages for the cocktails, as needed. But now, Jason said, "I think we can start breathing again."

Recovery from near death is not a linear process. It might seem to go without saying that being as sick as Tom was for as long as he was, and enduring the physical and psychological suffering that went with it, would be traumatic. But with all the life-or-death medical crises from day to day, the ongoing trauma became more of an existential backdrop than a clinical concern that was acknowledged and addressed. A number of factors can contribute to delirium: metabolic factors, depression, medication, or an uncontrolled infection. Mayhem in the body expresses itself in the mind. Much later, we would learn about Posttraumatic Stress Disorder (PTSD) and how seriously ill patients and their families, especially those whose experience includes time in an ICU, may develop emotional triggers that continue to set off anxiety, depression, or a range of stress-related behaviors. Untreated, it may come and go in episodes that everyone tries to tough out until the "mood" passes, but ignoring it doesn't really make it go away.

No wonder. As Tom described it to me one day, in his professor of psychiatry mode, the brain is made up of about 20 trillion neurons that are wired to work in harmony, bundled to process different tasks involving different parts of the brain—cognition, emotion, memory. Ordinarily, big emotions like love or fear would flip the switch on some bundles, while reason and memory would use others to keep things steady. With PTSD, he explained, whenever he thought *about* a hallucination, it was like looking at a memory circuit in his brain. But as the memory flooded back "it was as if all of the 20 billion neurons in my brain were bundled into a single high-voltage wire. When I touched that wire there was nothing else but emotion." No capacity to rationally differentiate between the past trauma and present moment, no way to keep things steady. Trying to forget the memories only left the tripwire hidden for repeated high-voltage shocks. With help, he was able to process the trauma differently, hold it as memory, but without the high-voltage wire.

Some days, Tom was depressed; other days, elated. Mostly, he seemed grateful to be alive and was filled with a new sense of awe about everything, no matter how mundane. He had to relearn how to do the simplest of tasks, like brushing his teeth and combing his hair. He played cards with Carly and thumbed through Sibley's *Field Guide to Birds of Western North America* with Frances like he'd done when the girls had been young. More often than not when I arrived at the hospital, I would find one or both of the girls perched on the side of his bed while they watched old movies. One afternoon, as they left and were out of earshot, he complained—about himself.

"Look at me," he said darkly. He held out his arms and stared at them in shock. I tried hard to hide my own. Once his kidney function had finally started to improve, the docs had started pumping him full of diuretics—the ones my grandma used to call "water pills"—that prevented the body from absorbing too much salt. As the extra fluids began to be siphoned off, he'd turned from a bloated pufferfish to a deflated balloon.

His forearms, which had been as thick as tree trunks, were now matchsticks. Folds of skin under his chin sagged and wobbled when he moved his head. The skin on his legs drooped like Granny's pantyhose.

"I'm an *old man*!" he said angrily. "I was playing cards with the girls, and I dropped some like a two-year-old because my hands won't close all the way!"

We were all so grateful he was alive, so relieved to even be able to see him conscious and conversant, that his appearance seemed secondary. But it was true: the robust Tom who'd taken steps two at a time up the face of the Red Pyramid had wasted away. There was a medical word for that: cachexia. Extreme weight loss and muscle wasting, commonly associated with malnutrition, debilitating disease... and dying. I remembered the first time I'd ever seen someone who looked this way. It was at Casey House, the AIDS hospice where I'd volunteered during the earliest days of the epidemic. My husband had become a walking skeleton. Or a living one, anyway. Walking would have to wait, but whatever else had wasted away, that bit of stubborn grit remained.

"I didn't make it this far to give up now," he said.

No retreat.

I bought him some wrist and ankle weights, and he dutifully exercised with them to regain his muscle strength, while I hummed the title soundtrack from the Stallone movie *Rocky*.

The day I wheeled him outside for the first time, a month after phage therapy began, it felt like we were going to prom. I brought along his favorite hat—a fedora that my parents had given him—and plopped it on his head, then wrapped a blanket around him. The nurse's aide, Carmen, helped me place two pillows under his butt, or what was left of it.

"Here we go, honey..." I whispered as I released the brakes on the wheelchair and pushed him out the double doors of the TICU. Carmen pushed the IV pole, as Marilyn, the charge nurse, looked on approvingly.

"Be back in fifteen minutes," she clucked like a chaperone.

"Or I'll turn into a pumpkin?" Tom retorted.

The look on Tom's face as I wheeled him past the palm trees in the lobby and the dapper doorman in the double-breasted uniform, out into the San Diego sun, was priceless. He told me later that when he took his first breath of truly fresh air, every molecule bombarded his senses. Cherry blossoms. Fresh cut grass. Espresso from the coffee wagon. He wept when he heard the familiar sound of a song sparrow calling for its mate: *Madge-Madge-Madge, put-on-your-tea-kettle-ettle-ettle.* This passionate ornithologist, a man whose ear was more attuned to birdsong than the chatter of his own species, had not heard a bird in nearly five months.

Little by little, though, the life-support accessories began to disappear, a reassuring sign. Tom's *ka* was making a comeback. A week later, he took his first baby steps in months, with Amy and the lift team supporting him on either side. A new CT showed that his pancreatic pseudocyst was no longer the size of a football, "but maybe a softball" said Tom Savides, the GI doc. On May 1, the cardiac monitor was unhooked and wheeled out of Bed 11.

"Good riddance!" Tom declared, with a note of finality and a ceremonial salute.

One afternoon in mid-May, six months since Tom had first fallen ill, Chip stood next to Tom's bed, perusing the lab report on his most recent blood test. He beamed at the opportunity to share good news.

"I am happy to confirm that we officially have the *Acinetobacter* on the run now," he said, smiling ear to ear. "In fact, two of the phages, including one that we added a few weeks ago, are synergistic with minocycline."

Tom looked at Chip in surprise. "Does that mean that the Super Killer phage is actually making the antibiotic work better?"

"Precisely," said Chip. "It's a perfect textbook example of synergy. Neither the antibiotic nor the phage were as effective on their own as they needed to be, but together—*bingo.* Stronger together. That is *the* definition of synergy." Limited studies with animals had suggested that the benefits of using phages with antibiotics together was greater than either one by itself, Chip said. Now Tom's case provided human evidence that antibiotic-phage synergy was not only possible but extremely promising.

Tom's immune system had rallied, too.

"Looks like Biswajit's *in vitro* results panned out *in vivo,*" Chip explained. "The synergistic effect made the bacteria more vulnerable to other bacteria-fighting agents, whether drugs or your immune system. Or maybe both."

The scientific literature was still thin on this front, but it was an area of urgent interest now in the fight against superbugs, he said. Researchers were searching through subtleties in drugs' interactions with drugs and with bacteria—and now phages could be part of that equation.

The possibilities were exciting, as Chip described them. This was just his kind of cutting-edge challenge, working with new data to save one patient, then pushing for new clinical trials for the data to save many more.

"From the phage therapy perspective, the serial samples we drew last week have provided some of the best phage PK data generated to date," he said, referring to the pharmokinetic activity, how the body absorbs, metabolizes, distributes, and eventually disposes of drugs or, in this case, phage. The data also confirmed the phage's continued synergy with the antibiotics, boosting the drugs' effectiveness. And the bold guess regarding dosing—how much and how often to administer the phage infusions—also appeared to have been the right guess to anticipate the complex interactions that would follow the phage infusions.

"There are still quite a few details to sort out," he said, cautious but not confounded. "But an ongoing presence of bacteria that shows some resistance to the phages means that the phages are still taking on the *Acinetobacter*."

Things could have easily turned out differently. Had we given up on phage therapy that roller-coaster Sunday morning when Tom's blood pressure had dropped suddenly because of the *Bacteroides* infection, Tom would not have survived, and we'd be just another anecdote in the strange history of phage therapy. The flip side of that would be to stop with a simple conclusion about the phage therapy success without learning more about all the contributing factors.

Whatever emerged in the new data from Tom's case, Chip said, the fact that it had been documented so carefully was a clinical coup worth celebrating.

"Although phage therapy has been around a long time, with Tom's case, we overcame a lot of the obstacles that have been preventing it from becoming more widely used to treat multi-drug-resistant bacterial infections outside of Georgia, Russia, and Poland."

"How so?" asked Tom, wide-eyed.

I piped up and told them about my readings of medical historian Bill Summers's takedown of the politics, prejudice, and skewed view of scientific evidence that had shut the door on phage therapy research in the US more than fifty years ago. Chip added that Carl Merril had expressed frustration at those who had dismissed phage therapy's potential on the basis of predictable bacterial resistance, choosing to ignore that phage could actually *respond* to resistance through mutations to counter it—the Darwinian dance. That was something drugs simply could not do.

"The FDA is doing cartwheels," Chip said. "This is the kind of data they need to start considering a new regulatory model so that phage therapy can hopefully be offered more easily to other patients someday, without having to secure an eIND for every phage cocktail."

"That's great!" Tom and I cheered in unison.

"I'm sure that's going to take some time," Chip continued, "but based on the success of Tom's case, the Navy is ramping up their phage program. They see the potential for offering phage therapy on demand, ideally using an ever-expanding phage bank. By partnering with industry, they hope to offer phage therapy to more civilians without some of the hoops we had to go through. And with your permission, Tom, we'll publish your case report to move the field forward."

Tom was all in. "Absolutely!"

I snorted in mock derision. "You're the only scientist I know who generated the most important publication in his career while lying in a coma."

"Speaking of brainstorms," Tom said, half musing but excited in the gleeful way he gets over *aha* moments in research. "Am I an *n* of one?

An *n* of one is scientist lingo meaning the only patient in a clinical trial.

"Was I the first person to get phage therapy?"

I shook my head no, and realized—again—that Tom had little to no recollection of all the stuff I'd been sharing with him about this all along. The history, the science, the bold guesses—and the big risks—that were part of his one-man miracle. He'd always teased me about my habit of launching into professorial mode with a full-blown lecture at any opportunity, so I tried to just stick to his question.

"Not exactly. Félix d'Hérelle was the true discoverer of bacteriophages one hundred years ago. He was also the first one to try phage therapy. Starting with himself..."

"My kind of guy!" Tom interrupted, marveling at his new invisible friend Félix.

"Speaking of your kind of guys," Chip chimed in as he pulled out his cell phone, searched for a moment before he found what he was looking for and passed the phone to Tom and me. There was a grainy black and white triptych—of phages. I'd looked at so many phages online that this seemed like another murky mug shot of the classic variety. With their angular geodesic heads and spindly legs, they really did look like alien spiders, or lunar landers, alongside the pudgy podophage. Or "pottyphage," as Tom called it, no disrespect intended.

"Meet your new BFFs," Chip said. "These are Theron's electron micrographs of your phages. And here's another one," he went on, scrolling through his smartphone. "This one's a scanning electron micrograph courtesy of one of Theron's colleagues, showing the Navy phages attacking your Iraqibacter."

Tom was instantly transfixed. Imagine being kidnapped and held hostage for six months, near death the whole time, and then meeting the superheroes who saved your life, heroes invisible to the naked eye. I was pretty wide-eyed myself. They'd saved our family, too.

"Monsters from the Black Lagoon!" Tom marveled, the pantheon of sci-fi mutants never far from his mind.

"Yeah," I mused, "but at least they're *our* monsters."

We soon saw just how beautiful these micro-monsters were. Forest and Anca, the dynamic duo from SDSU whose lab had stepped in at the eleventh hour to repurify the Texas phage cocktail, stopped by one afternoon bearing a gift. It was a copy of the lavishly illustrated *Life in Our Phage World: A Centennial Field Guide to the Earth's Most Diverse Inhabitants*, which Forest had co-written with several collaborators, all experts in various dimensions of phage ecology and history. A series of pen-and-ink phage portraits by illustrator Ben Darby caught our eye. As podophage family resemblance goes, enterobacteria phage T7 was a dead ringer for our Super Killer. Roundish and stout, with compact appendages. Practically cuddly. Looks can be deceiving.

"That puffy beachball is our Super Killer?!" Tom chortled. The cute factor was undeniable.

One of T7's traits, as described in the *Field Guide*, is that it "delivers its genome in a slow, highly-controlled fashion, thereby evading a host defense." Bingo. That had proved to be one of the superpowers of our Super Killer. And its habitat: mammalian intestines and sewage. *Delish.*

The pantheon of other phages is wildly diverse, and not even a drop in the bucket of the perhaps tens of trillions of types of phages roaming our mammalian guts, oceans, and sewers. Streptococcus phage 2972 is a long, lean, elegant slip of a thing with a delicate tassel for legs, known for being able to evade CRISPR defenses. Bacillus phage Φ29 and bacillus phage PZA look like something out of *Game of Thrones*, spiked bludgeons, no legs at all, adept at evasive strategies. Pseudomonas phage Φ6, partial to parasitizing plant pathogens, resembles a croquet ball covered with soccer cleats.

Phages have been called the dark matter of the virosphere, reflecting how little we know about them and the estimated two billion pieces of genetic code they carry around. But unlike the sci-fi drama, Forest's book—and our own experience—brought that teeming world into a brighter light. Forest had published the *Field Guide* to commemorate the one hundredth anniversary of phage biology, but he dedicated the book "to the second century of phage explorers." Tom had earned his explorer's badge the hard way.

Forest and Anca glowed when they saw Tom awake and alert. From the start, when they'd delivered the first round of Texas phage, they had both checked in on him periodically when one or the other would swing by to pick up Tom's newest isolate samples or drop off another round of purified phage prep.

"The 'before' and 'after' is pretty dramatic," Anca told Tom.

In this Darwinian ball, the phages were having the last dance. A victory dance.

28

THE BUDDHA'S GIFT

August 2016

When the day finally came, I kept pinching myself. Were we dreaming? After being hospitalized for nearly nine months, Tom was finally discharged from the hospital on August 12, 2016, and came home. Day 259 from the start of his illness in Luxor. He'd been admitted to Thornton, via Luxor and Frankfurt, on December 12, 2015. It made for an odd anniversary.

We'd already started celebrating the small milestones. A couple weeks earlier, when the last of his drains had been removed, we liked the new direction.

"Nice to have you guys removing drains instead of adding them, for a change," I said to Tom Savides, when he'd done the deed. After the procedure, my Tom pulled up his hospital gown to show me the holes where the last two drains had been.

"Looks like Dracula missed," I teased, adjusting the Velcro strap on his wrist weights to keep them from chafing. "But I guess we showed the docs that a Hail Mary pass *is* possible when the quarterback is blindfolded with less than a minute left in the game."

"That makes you the quarterback," Tom quipped.

"And Chip, and—"

"Time to get T-shirts for the team," Tom said brightly. Only this one can say WE BEAT IRAQIBACTER!"

We had. The two-part phage therapy, begun on March 15, had continued for eight weeks, concluding on May 12, when it appeared that *A. baumannii*, while present in some blood workups, was at a level that Tom's immune system was able to manage. Until then, the second-generation phage cocktail continued to prey on the *A. baumannii* and synergize with one of his antibiotics to make it more effective. This was the one-two punch Tom's immune system had needed to catch a break and return to action.

Eight weeks. Eight weeks that had put the brakes on a lethal infection that had raged for nearly four months before, an antibiotic-resistant superbug that moved like a wrecking ball through Tom's body before it was finally stopped. We would never know for sure exactly what triggered the gallstone pancreatitis or the pseudocyst, or when, and we would never know precisely where Tom picked up the *A. baumannii*. But from the day Dr. Zeuzem in Frankfurt had identified the football-size pseudocyst and the presence of multi-drug-resistant *A. baumannii*, the big-gun antibiotics had taken their own untold toll on Tom's system with negligible effect on the infection. At least fifteen antibiotics, most of them tried in the first six weeks of his illness, had failed to stop Tom's *A. baumannii*. And the uncontrolled infection had contributed to the cascade of other complications that pushed Tom so close to dying. We couldn't help but think now how this personalized phage therapy—had it been available and done earlier—could have helped so much sooner, slowed the extreme deterioration Tom suffered, and made for a shorter and smoother recovery.

This was what Carl Merril had envisioned decades earlier, and what was on his mind when he and I struck up a pen-pal relationship once it was clear that Tom was finally on the mend.

"I'm not sure who feels better, Tom or me," he wrote. He and Biswajit had struggled so hard and for so long to move phage therapy forward in those early years, only to come to the seeming dead end of retirement for him, and Biswajit's move was on to other work. Like so many of the scientists I'd come to know on this journey, Carl shared that Tom's case had profound meaning for him that reached back to his youth. He traced his earliest passion for science to summers with his grandfather, an accomplished electrical engineer, who taught him to read, physics books in particular, along with mathematics through calculus and matrix algebra. His grandfather's love of sports car design was also infectious. Disassembling, and more important, reassembling, those high-performance engines, and driving the cars, had become a serious pastime for Carl, honing the same intensity he brought to his career as a scientist. Plus it offered a release.

"With the cars, I at least had the illusion of control, and if something broke, I could fix it," Carl wrote. "This was a degree of control that I could not achieve in either medicine or the laboratory, so in a way, the cars offered an escape to a world where I had more control." In retirement, Carl's penchant for fast sports cars and building models for his grandkids drew smiles from others, but for him, the high-pressure, high-stakes challenge of Tom's case, and the opportunity to contribute as he did, had brought the thrill of it all full circle. A kind of healing.

Caring for Tom at home was no simple matter. I arranged for regular visits from mobile nurses, physical therapists, occupational therapists, and aides. We had a hospital bed set up in the living room. Tom's childhood friend Allen built temporary wheelchair ramps and installed rails in the bathroom. Although he came home in a wheelchair, Tom graduated to a cane within a few weeks. A month later, he was walking without it. Day by day, he improved, but recovery took a long time.

"Can't say we weren't warned," I said to Tom one day. "Do you remember a few weeks after you got to Thornton, Tom Savides told us that the rule of thumb was that for every week of hospitalization, it takes five weeks to recover?"

"I don't even want to do the math on nine months," Tom replied with a deep sigh.

On the bright side, Carly and Danny moved from San Francisco to

San Diego that summer. Carly visited regularly, and she and Tom went for walks and played boisterous games of cribbage. We enjoyed frequent visits with our other kids, Suzi, my parents, and a steady stream of friends, students, and staff.

I wish I had videotaped the moment Tom and our Maine coon cat, Newtie, were reunited. A few months earlier, Newt had had a near-death experience himself after I put a permanent stop to him stealing the kittens' food. I was so busy at the hospital, I hadn't noticed that Newt had stopped eating altogether. He suffered kidney failure and ended up on a feeding tube just like his master. Over the next year, he and Tom would nurse each other back to health, Newt curled up in the crook of Tom's elbow, gently snoring.

One afternoon in September, Chip brought a special visitor out to see us. When I opened the door, there, in Navy fatigues, stood a youthful Tom Cruise lookalike, Lt. Commander Theron Hamilton from the US Navy BDRC lab that had supervised the preparation of Tom's two IV phage cocktails. "Amazing Theron," as we'd come to call him, had been quick to respond to Chip's initial SOS for phages and then had ushered the formal request through the required ranks for approval. Time had only deepened our awe that Theron had been able to so nimbly navigate the military bureaucracy, and that he and Biswajit hadn't hesitated to turn their lives—and their lab—upside down to launch the offensive against the *A. baumannii*. This was one naval battle that had been fought and won in the lab.

We'd seen Chip since Tom's discharge, but when he and Theron walked through the door, it was instant euphoria all around. Tom and I were overwhelmed to meet Theron in person. His decisiveness early on to jumpstart the Navy process, press the admirals for approval, and then to support Biswajit and his lab at every turn was inspiring. At a scientific workshop where he presented Tom's case, Biswajit later described his boss as "brave." I'm not sure that kind of courage—which clearly runs deep in the ranks of Navy scientists—gets celebrated in the ad campaigns, but it should. Theron, long a steely smart superhero in our minds, turned out to be a warm, tender-hearted one, too. I thought I saw all three guys with tears in their eyes at first, but in a flash, they were all laughing and back-slapping.

Chip was recovering, too. The old familiar spark was back. Throughout the ordeal, he'd never flagged, never lacked the energy or attention for his round-the-clock doctoring. Or if he had, he had not shown it to us. And he'd never lost his sense of humor. But when he was tense, his humor had a dark edge, and to see him relaxed now, we realized we'd all been living in the shadows for a very long time.

Soon after Tom came home, we began to talk—really talk—about what had happened. We realized that we had gone through the same ordeal but had different experiences. Although we had now been married twelve years and spent almost every day of his hospitalization together, nothing prepared us for how we would feel when we learned about what the other had seen, felt, and thought through the last nine months.

"Remember the time in the ICU in Frankfurt when you demanded a smorgasbord of food, and then promptly threw it all up?" I asked Tom as I helped him sit down at the breakfast table one morning a few weeks after he came home. "You spewed black projectile everywhere. I told you if your head had spun around 180 degrees, you could have starred in the next *Exorcist* movie!"

Tom stared. "I did?" He ignored his scrambled eggs and gazed off into space, tugging at the tendrils of a memory like a gnarled piece of yarn.

"I was a Buddha," he murmured. "I was sitting in a lotus position, feeling absolutely beatific, on another plane." He extended his arms, palms up, and closed his eyes before he continued. "I was so at peace, I wanted to bestow upon the world a gift, so I opened my mouth, and all of these silver ribbons of tinfoil swirled out. They were so beautiful as they fell to the ground, how the flecks of light made them sparkle. Everyone was running around, picking them up; they were so happy."

He wasn't joking. I gaped, then snorted. "A gift?! Not bloody likely! It was a *mess*!"

We both erupted in maniacal gales of laughter, then shuddering gasps that morphed into tears. We held each other and wept, one of the first of many such moments we would have over the next few years, as we processed what had happened.

"I'm so sorry," Tom whispered, burying his face in my neck.

"What are you sorry for? I'm sorry you had to go through all that!" I clung to Tom's shoulder and then pulled away, looking at my husband with new eyes. He was a changed man, physically, mentally, and spiritually. Try as I might, I would never really know what had gone on inside his mind and his body all those months.

Years before he fell ill, Tom had started collecting Buddha statues. He'd had a particular fascination with the Starving Buddha, which had a skeletal appearance, depicting the suffering that Buddha experienced through an extreme fasting period prior to his enlightenment. Nirvana might have to wait, but Tom had certainly transcended states of ordinary consciousness in his months-long "fast," and maybe stumbled onto his own Starving Buddha, the one within.

"We have to get to know each other again," I whispered. "And it starts with understanding what happened."

"Write it down," Tom told me. "I want to tell you about my hallucinations, my dreams."

Thus began our journaling. On sabbatical from the university, I would rise every morning at five a.m. and write. Furiously. Each day, I would consult my entries from Facebook—fifty-two pages worth—and write out my version of our story, consulting at times with Carly and Frances, or Chip and Davey for the medical details. Tom would rise later, and while sipping from his mug of Peet's, he would seemingly enter a trance, dictating his hallucinations in vivid detail from the searing memory of them—memories that he had struggled at times to forget.

We shared some of the entries with my parents, Cameron, and the girls, and read some aloud to Jill's daughters. Now that Tom was out of the woods, everyone started to ask questions about things they'd wondered about but were too polite or afraid to ask.

"What's a superbug?" asked my twelve-year-old niece, Morgan, after I'd read our first entry to her and her sister, Rylie.

"Will I almost die too, if I catch one?" asked Rylie, who had just turned fifteen.

"Not necessarily," I explained. "But infections that used to be easily cured are getting harder and harder to treat, which is why Tom and I wanted to be sure that no matter what happened to him with this treatment, the data could be used to advance research so it could help other people."

Cameron was more direct with his questions over dinner with Tom and me later that year, during our first trip to Vancouver since Tom was discharged.

"What was it like being in a coma?" he asked, shoveling spaghetti into his mouth.

Tom tried to explain, then immediately got choked up. Although he was improving physically, he was increasingly becoming overwhelmed with the terror of his hallucinations and was having nightmares.

"I can't stop thinking about my dreams," he confided. "They're so lifelike. My brain goes right back to thinking I'm in the TICU."

So did mine.

"It's frightening not knowing what is real and unreal," Tom said. "To not be able to trust your brain."

Tom was relieved to finally be able to talk with us about his experience. He now distinguished between the *disease*—the *Acinetobacter* infection and all the rest—which was his alone, and the *illness*, which was what everybody else around him experienced.

"Every medical case is lived twice: once in the wards and once in memory," wrote Siddhartha Mukherjee, physician and author, paraphrasing the writer Viet Thanh Nguyen.

For a couple or a family, every medical case is lived twice more: alone and together. Each of us had our own version of the illness experience as it affected us individually. Our shared version as a family was a patchwork of pieces that came together more slowly with time and conversations. It was another kind of healing.

29

GRAND ROUNDS

One year later

Chip Schooley was in his element. Or at least one of them. He was as at home in this cavernous UCSD lecture hall as he was bedside with a patient, consulting from Mozambique, or orienting a new medical resident to the inner labyrinth of the sprawling medical center and adjacent university grounds. Even as he'd prepared for this milestone grand rounds event, presenting Tom's case for the first time to our UCSD clinical community, he had a ticket in his pocket for a flight to DC that evening. He planned to meet the next day with NIH officials there to discuss funding for clinical research trials so that phage therapy could one day be used to do the most good for the most people.

It was little more than a year from the day Tom had left Thornton Hospital in a wheelchair to come home, and this morning we'd parked and he'd ambled up the grassy knoll to the Liebow Auditorium for grand rounds, no assist necessary. We were there to mark the moment and join Chip at the lectern for the start and close of the clinical presentation to share a little from our experience as The Patient and The Patient's Wife.

The walk up the hill had been easy enough, but we both had to catch our breath at the sight of all these people, so many of them having had a hand in saving Tom's life. Randy Taplitz, Kim Kerr, Sharon Reed, Atul Malhotra, Eric Scholten, and several of the TICU nurses as well as Connie Benson, who we knew had partnered every step of the way as Chip had turned their living room into the war room to strategize moves to contain and kill the *Acinetobacter.* My friend and neighbor Liz Greer came too, taking time off from her job as the pacemaker nurse at the nearby VA. And there, some rows back, looking for all the world like a shaggy surfer who had just happened by, was Forest Rohwer, whose rare expertise in phage purification and his instant willingness to help had made it possible to get the Texas cocktail to Tom in time for the critical first strike.

Each of them knew intimately a piece of the picture that Chip was now about to make whole.

I thought of others on the phage therapy team who couldn't attend the presentation but were there in spirit: two labs in the crisis collaboration, one team at a university in College Station, Texas, the other at a restricted military lab in Maryland. Carl Merril would likely be chatting with his son Greg and walking Rocky, his steadfast border collie. And the administrative "trail angels" who had seen to it that the documentation required for regulatory and ethics review was done and delivered by impossible deadlines. All told, our threads of emails and texts—thousands of them over the past year—had translated science from bench to bedside with unheard-of speed. Each one of these people, and others we would never know, had been critical to pulling this off.

Our friend and clinical confidant Davey had to be out of town today, but he never felt far from us anymore. Tom pointed out, as he thought of the team effort and this particular duo of doctors: "Chip kept my body alive, but Davey kept my spirit alive."

The sense of something whole coming into view reached deeper still. During Tom's hospitalization, especially in some of the most intense or chaotic moments, I sometimes felt myself suddenly detached, witnessing the scene and myself in it as an observer. Not judging, just observing. That frantic bleary-eyed woman in the hoodie, racing after the doctors and Tom's gurney down the hall for the emergency procedure when his drain slipped? I watched her run. Watched her cry sometimes. Watched her dance by her husband's bed. Watched her flip into driven-scientist mode and take it too far at times, pushing too hard, snapping at people impatiently or angrily. The sense of fracture between my two selves—the loving caregiver and the pit bull scientist—was always there. Like the shattered pieces of colored glass in a kaleidoscope that just keeps turning, they had always felt fractious and incompatible. Today, as I watched Chip move through his presentation, Tom by my side and the spirit of so many with us, the kaleidoscope suddenly stopped churning and the pieces fell calmly into place. The wife and the scientist felt whole at last, and the tension of the twin imperatives, one to save Tom, the other to generate useful data even if he died—came to a rest, too. *Mission accomplished.*

This grand rounds presentation was the way that wholeness looks in clinical medicine, dissecting the symptoms, distilling the case, and passing the lessons learned to other clinicians and students. That's no mean feat with a medical record 3,981 pages long, encompassing nearly nine months.

Biswajit had presented Tom's case to an equally enthusiastic audience at the Pasteur Institute in Paris in April for the Centennial Celebration of Bacteriophage Research, reaching an international community of scientists and clinicians that included Félix d'Hérelle's great-grandson, Dr. Hubert Mazure. And just two months earlier, the FDA had hosted a phage workshop open to the public in Bethesda, highlighting both the history and advances in phage therapy. The program included multiple presentations by those involved in Tom's case, including Biswajit and other Navy scientists, Ry and Jason from Texas A&M, and Chip, all reporting on different aspects of the case. Tom's case report was slated for publication in an upcoming issue of *Antimicrobial Agents and Chemotherapy*, as well as a commentary discussing the case in *The Lancet*, and an interview with Chip in the *Journal of the American Medical Association*. The Patterson case was now well documented, with an IV phage therapy protocol available globally to help others.

From grand rounds to the social media grapevine, those lessons were beginning to fill in some blanks with new evidence and possibility:

First, phage therapy warranted a fresh look as a potentially personalized treatment against multi-drug-resistant bacterial infections. More basic research and clinical trials were needed to advance the field, and if those studies were favorable, a new regulatory pathway was needed to bring it to scale.

Second, the medical establishment and funding agencies needed to overcome their implicit bias against the use of unconventional treatments like phage therapy, which was first studied prior to the dawn of molecular biology and thus had suffered from scientific and logistical limitations.

Third, Tom's case suggested that it was possible to co-administer phages with antibiotics to enhance the effects of both. Another case report from Yale that used phages to resensitize a different bacterium to antibiotics reinforced this possibility.

Finally, the international effort to save the life of a single individual was global health diplomacy in action. Urgent global health challenges require actors at multiple levels and creative ways to maneuver logistical hurdles that impede progress. One month after Tom was discharged from the hospital, a 2016 UN General Assembly report called for multisectorial and cross-sectorial efforts and engagement of all relevant sectors of society—human and veterinary medicine, agriculture, finance, environment, industry, and consumers—to tackle the global superbug crisis. The consensus statement was signed by all 193 countries. With millions of lives at stake due to the growing threat of antimicrobial resistance, this kind of global collaboration is essential.

That complex collaboration could appear deceptively simple in Tom's case because, in the end, all most people saw was that a very sick man was in the hospital for a very long time, recovered, and went home. Some people called it a medical miracle, and it was. But from our vantage point, we'd seen the constellation of coincidences and variables that could have gone so differently, the teamwork across disciplines and miles that could have easily broken down but didn't, the just-in-time delivery of the people and resources with a role in this, and we saw each one as a singular miracle in itself.

The FDA's Cara Fiore had marveled at the astonishing synchronicity of it all. She'd told Chip that to have obtained the FDA's approval of phage therapy for a clinical use, they'd needed to have a patient with a multi-drug-resistant bacterial infection who was dying, a caregiver who agreed to pursue phage therapy, a doctor and a medical institution who would cut through the mountain of red tape, and a network of phage researchers who were willing to go flat-out to launch a phage hunt to identify, purify, and prepare a phage cocktail to the FDA's rigorous standard. They couldn't have dreamed of a better situation than the one Chip handed them. It occurred to me that there couldn't have been a better person to lead the phage protocol than Chip, either. He was exactly the kind of person who could take on the old-timers in the medical establishment who had been influenced by the negative reviews about phages from the 1930s. Not only that, he was able to navigate the contemporary landscape to coordinate initiatives for rigorous, responsible, progressive phage therapy research and clinical studies; be a mentor in the medical field as it sought ways to help patients; and continue, as a physician himself, to practice the art and science of medicine on the cutting edge.

It struck me, especially from my perspective in global health, how Chip, and so many who had partnered with us through this odyssey, embodied the spirit and potential of medical science in the global age. Each in their own way had reached out to bring together the best from everyone and everything, everywhere. Chip told me later how, at the critical juncture when Tom's phage therapy was temporarily halted, he'd called a renowned expert on the subject, his longtime colleague and friend Charles Dinarello, a professor of medicine at University of Colorado. When Charles didn't answer the phone that Sunday morning, Chip had emailed him. Charles had answered within minutes from a Delta flight somewhere over Greenland. After a lengthy discussion, during which Charles offered to measure endotoxin in Tom's blood, Charles ultimately concluded that endotoxins weren't the cause of his septic shock episode. His reassurance had freed Chip to refocus on other possibilities.

So many times, this pickup team of doctors and scientists without borders had made the difference between life and death for Tom. I would never know them all, but I could feel the global village present with us at grand rounds.

Sir Isaac Newton famously said of his own accomplishments that if he could see further than others, it was because he had "stood on the shoulders of giants." We were acutely aware of the shoulders on which the success of Tom's phage therapy rested. Tom wasn't the first person to be treated and cured with phage therapy, not by a long shot. Beginning with Félix d'Hérelle a hundred years ago, scientists around the world had spent their careers on phage research and its clinical applications, beginning with the Republic of Georgia, Russia, Poland, and more recently, Belgium. In the US, phage researchers like Betty Kutter from Evergreen State College had used phage therapy topically to successfully treat diabetic toe ulcers. Félix and his contemporaries died before their work could be fully appreciated, but all those on the phage front then and today had supplied essential pieces of the knowledge and expertise that had saved Tom's life. Our "bold guess" to treat Tom intravenously with phage cocktails had paid off, but only because we had learned from the experiences of so many others—their successes and failures—and taken a calculated risk.

In truth, this was also a story about privilege. So many people who are dying of multi-drug-resistant bacterial infections don't have the connections and resources to call upon that we did. If phage therapy is shown efficacious in clinical trials, our goal is to help ensure that it is scaled up in lower- and middle-income countries that bear the largest burden of the superbug crisis.

As fortunate as Tom was to survive his battle against one of the world's most lethal superbugs, his was a worst-case scenario that showed just how fast we're losing that battle globally and how unprepared our US medical and healthcare systems are for the crisis. The plasmid that confers resistance to the antibiotic colistin was reported in China in November 2015—the same time that Tom got sick. Once it was discovered, the Chinese stopped using colistin in livestock. But resistance to it had already spread to thirty countries and five bacterial species before anyone noticed.

In Tom's case, when we were in Frankfurt waiting for the antibiotic resistance profile to come back on Tom's *A. baumannii* isolate, Chip said he'd be very surprised if it came back resistant to colistin, but by the time we got to San Diego two weeks later, Tom's *Acinetobacter* was fully resistant to it. That's how quickly these superbugs can develop resistance and catch us off guard.

The spread of colistin resistance was a failure on so many levels. Widespread use in livestock of a medically important antibiotic for humans was known to contribute to antibiotic resistance more than a decade ago. The lack of a surveillance system for detecting the emergence of colistin-resistant bacteria had hidden the growing threat for an unknown period. And the unavailability of rapid tests to diagnose bacterial infections and determine which antibiotic might work slowed any response time, giving superbugs an even greater head start. A 2016 report found that colistin was still being used in agriculture in countries like India, Vietnam, Russia, Mexico, Colombia, and Bolivia.

In the US, it wasn't until 2017 that the FDA banned use of medically important antibiotics to promote growth in livestock. Although antibiotics used for prevention in the US are now under the control of veterinarians, many of these are unlabeled growth promoters, which continue to be one of the main contributors to the spread of multi-drug-resistant bacteria. And banning use of the antibiotics for that purpose here hasn't kept some in the animal pharmaceutical industry from continuing to sell them as a growth enhancer in countries where regulation and enforcement has been more lax, exposing those consumers to greater risk and adding to the global factors contributing to antimicrobial resistance.

Experts now acknowledge that infectious diseases we thought we'd licked are making a comeback, and some infections that once were easily cured with antibiotics are almost untreatable now. What's more, many routine procedures like colonoscopies and common surgeries like joint replacements now present a greater risk of infection by antibiotic-resistant bacteria present in hospital settings and equipment.

There is no new family of antibiotics on the horizon that promises to rescue us as penicillin once did. However, there is precedent for innovative breakthroughs in science and medicine when disasters strike and our backs are up against the wall. The urgency of war led to a number of medical advances in World War II, penicillin topping the list. If not for the mounting pressure to find a cure for battlefield wounds, would penicillin have been tested and scaled up at the time? When the pressure is on and conventional methods fail, alternatives get more serious attention, and that is often how old ideas get recycled, new discoveries are made, and new cures are developed.

Desperation was what pushed me to action, too. And that sense of urgency paired with possibility had spurred the actions by others who found a way to take what was known and push forward, which eventually made Tom's one-man miracle possible. But with a bold guess comes risks, and the risk/benefit ratio is one that doctors take into account every day, as do the FDA, the ethics committees at hospitals and universities, and the loved ones who make the ultimate decision.

Tom and I, as well as Chip, Davey, and Connie, had witnessed this firsthand in the HIV epidemic. When there were no AIDS drugs available, researchers and clinicians pushed for clinical trials, whereas AIDS activists pressed for access to the experimental treatments. There were terrible clashes at first, and as a PhD student, I felt caught in the middle, seeing both sides. At the 1991 International AIDS Conference in San Francisco, I'd even participated in a "die-in" protest with ACT-UP activists outside the conference hall, when my researcher colleagues were inside, and faced off against police in riot gear as we all lay down on the pavement and had other activists outline our bodies in spray paint. My friends and even my PhD adviser, Randy, were dying, waiting for drugs that could save them or give them even a little more time, whereas researchers were saying it was too early, too risky, to allow patients access to drugs that could turn out to be toxic. In the end, there was room for both. Clinical trials went forward, but a pathway for compassionate use was created on a case-by-case basis, where the FDA, ethics committees, doctors, and patients could weigh the potential benefits versus the risks based on each patient's unique situation.

That was our hope when we arranged for Tom's phage therapy protocol to provide data in a way that could be useful to other scientists and physicians. We were aware that the phage cocktails themselves—and administering them intravenously in particular—posed risks and set aside drug safety standards that might never be acceptable for general use. But what tipped the scales for us was the possibility that Tom's case—even if it meant Tom's death—could contribute to scientific understanding of phage therapy and to the impetus for needed clinical research.

The surprise in store for us was that it happened so quickly. When Tom's case was first covered in the news in April 2017, after Biswajit presented it in Paris, our story went viral on social media.

Within a few days, I received the first call for help from a woman whose family member was suffering from a multi-drug-resistant *Acinetobacter* infection in China. That was soon followed by others with chronic urinary tract infections, lung infections associated with cystic fibrosis, and complications following surgeries, like sepsis. All had superbugs and all wanted phage therapy. Desperately. I turned to Chip, Theron, and Ry for help. Some patients died before we could obtain their samples to find matching phage. A few went to the Eliava Phage Center in Tbilisi. And in a growing number of cases, we were able to help.

Later that year, Belgian researchers working with Jean-Paul Pirnay and Maia Merabishvili reported on successful intravenous phage therapy to treat a patient with a life-threatening multi-drug-resistant infection. In early June, based on the success of Tom's case, phage therapy was used successfully to treat another patient at UCSD who was battling a lung infection following a double lung transplant. Tom and I met John Willson and his family one day when we were visiting the TICU nurses and doctors. As we stood bedside with several of John's family members donned in the all-too-familiar yellow gown and blue gloves, John's daughter Jolynn embraced us and held our hands.

"Thank you for your courage," she said. "Because of you, my dad is still alive."

However, the "shoulders of giants" are often those of the most vulnerable among us. During Tom's recovery, Chip was contacted about using a modified version of the protocol to help a two-year-old boy, working with Theron and Biswajit's phage lab at the Navy. The boy's parents decided to try phage therapy after being presented with Tom's case, and it was working; his infection cleared. Sadly, he died of his underlying heart problem, but his case added to the body of evidence that phage therapy is worth a closer look.

Just weeks after the grand rounds presentation, I received a call from the father of a young woman not much older than Cameron. Mark Smith asked if I could help his daughter Mallory obtain phage therapy. A cystic fibrosis patient, Mallory had recently had a double lung transplant after battling a chronic superbug infection with an uncommon but nasty bacteria, *Burkholderia cepacia*. The infection had come back and was now attacking her new lungs. Cystic fibrosis patients have a genetic mutation that means they can't clear mucus from their lungs, which breeds infection. They receive a lot of antibiotics, which, over time, fuels resistance.

Mallory's mom, Diane, sent me a photo of herself and Mark with Mallory from her hospital bed in Pittsburgh, where she had recently celebrated her twenty-fifth birthday. I pinned it to my screensaver for inspiration. Chip, Theron, Biswajit, Ry, and one of his colleagues at Texas A&M were all in, but a few were skeptical that we could find phages against *B. cepacia*. And *B. cepacia* phages were tricky. They tended to be temperate rather than lytic, meaning that even if they matched Mallory's superbug, they could integrate into the bacterial DNA—essentially merge with it—without killing it. Since there was no time to sequence the phages, Ry, in particular, was worried that these phages could carry harmful toxin or antibiotic resistance genes. Still, the Smiths were willing to try it if we could find phages to match. They had reached out to the only phage researcher they could find who had published on *B. cepacia*, Dr. Jon Dennis from Alberta, Canada. He too, was on board. But we would need a phage cocktail, which meant launching a wider net. It felt like déjà vu. *Could we do it all again?*

The PubMed database didn't identify any other researchers who studied *B. cepacia* phages. After racking my brain about how to find more phage researchers who could help, I turned to Twitter. Maybe we could crowdsource phages for Mallory. My tweet was retweeted 432 times and found several other phage researchers who stepped up to the plate. Mallory's parents had her bacterial isolate sent to their labs, which included Texas A&M and Adaptive Phage Therapeutics (APT), a new phage therapy startup in Maryland that Carl Merril and his son Greg had launched after witnessing the success of Tom's case. After working around the clock for several days, Dr. Dennis and APT, working with Biswajit and the Navy lab, both found matching phages. We were exhilarated; success appeared within reach. But before the phages could be purified and amplified, Mark called with terrible news. Mallory's condition had worsened considerably, and her doctors said they could only keep her alive for a few more days. Mark and Diane decided to take whatever phage preparation was available, even though it wasn't fully purified or amplified, and give it a try. A few small vials were raced to Pittsburgh. Mallory only received a few micro doses before her doctors decided that it was too late. When she was taken off life support, we were all crushed.

"When I first heard about phage therapy," Mark told me at the reception after Mallory's funeral, "I had an idea. I asked Mallory's doctors if we could treat her before she had her lung transplant, so we could clear her infection and give her new lungs a chance. But they had never heard of this treatment. They thought it was too risky and dismissed it. But I can't help but wonder if it could have worked."

We had an opportunity to try Mark's idea. In early 2018, Chip and Dr. Saima Aslam from UC San Diego treated a cystic fibrosis patient with phage therapy, a woman the same age Mallory would have been. But this time, she received phage therapy *before* she received a lung transplant, with the hopes that if her infection could be cleared, her new lungs would avoid reinfection. Her phage treatment worked beautifully, and she returned home to wait for her lung transplant.

We may have learned as much from Mallory's death as we have from Tom's survival. As Ry liked to say, "phage therapy 2.0 is on its way."

****

One morning in summer 2018, incredulous upon hearing news of the US federal government's plans to stop public reporting of hospital-acquired infections and to continue the use of preventive antibiotics in agribusiness, Tom and I sat, discouraged, before heading into work for the day.

"This superbug crisis is bigger than us," I told Tom. "Why aren't people paying attention?"

Tom watched me pensively. "Look at it this way," he said. "You're an infectious disease epidemiologist, right? But until Iraqibacter almost took me down, it wasn't like you were going around sounding the alarm bells about the superbug crisis."

He had a point. And by dosing Tom with Cipro in Egypt without a doctor's oversight, my misuse of antibiotics was part of the problem. I had been clueless in Frankfurt and even here, to truly comprehend what we were up against—what Tom *and* his medical team were up against.

I had always told my students that being ensconced in the ivory tower of academia could render scientists out of touch. If people like me who are supposed to know better are oblivious to the urgent threat superbugs pose to human civilization, how can we expect to inspire global action for change? As a species, are we ignoring the signs? Suffering from collective denial? Or do we just think that we humans are supreme beings that can outsmart any microbe?

A post-antibiotic era. That's how some of the world's top health officials, including former CDC director Tom Frieden, describe the global threat of antimicrobial resistance (AMR). By 2050, one person could die from a superbug infection every three seconds, making AMR a more immediate threat to humankind than climate change.

In 2018, the world's leading health agencies worked together to stop an Ebola outbreak that threatened to spiral out of control and announced that globally, more than half of people living with HIV were now receiving HIV antiretrovirals. But despite urgent calls for action, little progress has been made to curb the global spread of AMR. New strains of antibiotic-resistant bacteria continue to be reported along with the spread of "extremely" drug-resistant bacteria associated with diseases like tuberculosis, gonorrhea, and typhoid.

Healthcare systems and the pharmaceutical industry are on the hook, too. A recent CDC report found that in US outpatient clinics, antibiotics were inappropriately prescribed almost half the time. In 2018, the first detailed analysis of pharma action against AMR found that of twenty-eight antibiotic candidates in late stages of clinical development, only two had plans to ensure that the product will be used wisely. Less than half of companies with antibiotics on the market are involved in AMR surveillance, and of eighteen major antibiotics manufacturers, just eight had set limits on antibiotic wastewater discharge. Meanwhile, the death toll from antimicrobial resistance in the US alone has been grossly underestimated in past analyses, and is now estimated to have been more than 150,000 in 2010—a seven-fold increase over previous reports. Roughly one-fifth of infections occuring in Europe, North America, and Australia are believed to be due to antibiotic-resistant bacteria.

In one sense, Tom's case was a singular one. But in truth, what happened to Tom could happen to anyone who acquires a multi-drug-resistant bacterial infection, and the chances of that are rising. We can't afford to be blindsided. Bacteria are evolving much faster than our ability to develop new antibiotics. Even if new antibiotics are found and phage therapy is found to be efficacious and brought to scale, it's naïve to think that we can simply treat ourselves out of the superbug crisis. We need to be proactive, not reactive.

What we did to save Tom was a collective effort, drawing from expertise past and present, and from corners of the world that have been overlooked. The "not alone" experience I'd first felt when phage researchers came to Tom's rescue and when family and friends helped us keep our hope alive—that energy and love came full circle as we faced forward now. With it, came the sense of "something larger" at work, not only at a personal level, but globally.

I keep thinking of the image of embers burning in a fire. So many times over the past hundred years, people have tried to light the fire under phage therapy, and the sparks went out. But this time, one little ember lit the flame. We don't know whether phage therapy will be shown efficacious in clinical trials, but it deserves rigorous evaluation. At the very least, if it can render some failing antibiotics useful again, that in and of itself is a game changer. Our work ahead is to get clinical trials of phage therapy underway, all the while helping other compassionate use cases. We may not have the conclusive evidence yet, but we do have precisely what is needed to carry the work forward. We now have evidence-based hope.

EPILOGUE

November 2018

> There is always hope in life, because there is always hope in science.
>
> —*Françoise Barré-Sinoussi*

Tom and I returned to work in April 2017, exactly a year after he was taken off life support. It's been anything but business as usual. Tom's recovery was multifaceted and included both physical and psychological healing. He had short-term memory problems at first—struggling to find a word, for example—but that cleared up. He still suffers from persistent numbness in his feet, possibly due to the gorilla-cillins he took or his worsening diabetes, since the infection destroyed one-third of his pancreas. After so many episodes of septic shock, his heart had taken a beating, so he was put on a fistful of prescription meds for mild congestive heart failure. He has trouble putting on weight, too. We could manage all that, but about four months after leaving the hospital, it was clear that Tom's PTSD wasn't going to go away on its own.

I was struggling, too. Instead of flashbacks to hallucinations, I seemed to have some kind of phantom limb syndrome, feeling the tug of emotions that were part of the nightmare we'd all endured—and supposedly moved past. The illness was over, but my reactions to it were not. I would overreact to the smallest hiccup. When Tom fell and skinned his knee, barfed up his breakfast, or once when I waited too long for a taxi, I went into full panic mode, adrenaline pumping. Biologists have a name for the physiologic stress response: fight or flight. It's the survival response that's triggered whenever we feel threatened. PTSD triggers similar deep wiring in response to a past trauma or terrifying event. In the context of hospitalization in intensive care, the disorder is called post–intensive care syndrome, or PICS.

PICS refers to short- and long-term cognitive, mental, and/or physical health problems following a critical illness, with its hallmark being symptoms that get worse after physical recovery. Government stats suggest that up to three-quarters of ICU patients may experience cognitive impairment, and up to 60 percent have PTSD. With 5 million people receiving care in the ICU per year in the US alone, of whom more than 750,000 require mechanical ventilation, as Tom did, that's a sizable at-risk population and an identifiable condition that's been largely overlooked. A year later, NIH would describe PICS as a public health burden "due to the associated neuropsychological and functional disability," but with scant attention to it previously, the agency concluded, "its exact prevalence remains unknown."

What we didn't know is that families of ICU survivors can experience PICS, too. That explained my own hyperresponsive symptoms. Once we recognized all of this as an identifiable problem—a problem that, however difficult, was treatable—we were relieved to discover there were people who could help us.

These health issues might seem tangential to the superbug story, but they are not. When multi-drug-resistant infections land someone in the hospital, prolong their stay, and present serious additional complications that they and their families must confront, those are the reality of the post-antibiotic world. That's the reality we all need to wake up to.

After the success of Tom's case, Chip and I continue to be approached by desperate patients, almost daily. He and other infectious disease physicians at UC San Diego have successfully treated five other patients with intravenous phage therapy and consulted on a growing number of other phage therapy cases across the US and internationally. Due in part to these efforts, phage therapy is being discussed more openly by researchers, biotech companies, infectious disease physicians, and patients with superbugs who have run out of antibiotic options. In July 2018, Pradeep Khosla, the UC San Diego chancellor, provided Chip and me with $1.2 million in seed funding over three years to launch the Center for Innovative Phage Applications and Therapeutics (IPATH), the first phage therapy center in North America. The day it opened, *Science* magazine published a commentary in which a respected microbiologist and physician was cited saying that IPATH was "a game changer for the field." To my delight, I received a congratulatory email from Dr. Hubert Mazure, the great-grandson of Félix d'Hérelle.

In the year ahead, a report of the first patient to undergo successful treatment with a genetically modified phage is slated for publication. New clinical trials of phage therapy are expected to be underway for patients battling superbug infections associated with cystic fibrosis and implanted medical devices. It will probably be a few years before the data are in, but if these trials show that phage therapy is as efficacious as antibiotics, the FDA will have what they need to make it more widely available. Some researchers dream of one day being able to use phage therapy to "groom" the gut microbiome to weed out harmful bacteria and promote healthy ones.

It's thrilling to play a role in moving phage therapy forward into the twenty-first century, but when your professional and personal lives collide as they did for Tom and me, it changes you forever. Total strangers helped us in our time of need, and we have broadened our efforts to give back, including our decision to write this book.

As a couple, we don't take any day for granted, and we don't sweat the small stuff. Just the other day, Tom took his first swim in the Pacific Ocean since his recovery. When he disappeared under a wave, I called out in a panic to Carly, who was closer to shore: "Where's your dad?" I could see her grinning as she pointed to a silver head that bobbed up beside her. "Here, silly! He just rode a wave in!"

Last year, Tom and I started traveling internationally again, including to Africa (over Thanksgiving, of course). Tom has a new bucket list, and Luxor's Valley of the Kings is still on it.

I've pondered what Robert had told me when Tom fell ill, that I had lived through all the experiences and acquired all the skills to deal with the challenges ahead. All the pieces were there, I just had to look harder to see it and to believe in my ability to get through the challenges each day brought. I believe that to be true for many of us.

Tom has reflected deeply on his near-death experiences. If you ask him what he thinks about the three questions the wise men asked him in his desert hallucination, he'll talk your ear off, but will inevitably say, "A wise person once said, the two most important days in your life are the day you were born and the day you find out why." He's still one hundred pounds lighter, but one thing Tom hasn't lost is his sense of humor. In July 2018, when he was promoted to Distinguished Professor in the Department of Psychiatry, he received a round of applause from his peers. As we toasted his accomplishment over a glass of wine that night, Tom raised his glass and said, "Here's to the 'Nearly Extinguished' Professor'!"

We're often asked how our work as HIV researchers prepared us in any way for our year in the grip of a superbug. We've mentioned elsewhere in this book how our work with the people and the issues and the science of HIV/AIDS has informed our lives deeply in many ways. Perhaps one of the most relevant, as it affects us all, is something that the documentary filmmaker Janet Tobias recently said to us. HIV researchers and activists believe that the impossible is possible—that it is possible to stop a global pandemic in its tracks, she said.

It's true. Tom, me, Chip, Connie, and Davey, all of us came from the testing grounds of the HIV epidemic. In the context of our battle against Tom's antibiotic-resistant *Acinetobacter*, for Chip and me, it meant that we never gave up hope that our bold guess would work. For Tom, it was about being resilient, like HIV's long-term survivors. Choosing to die is easy, he tells me; choosing to live when you are in such incredible pain is much, much harder.

Not everyone gets to choose. We know that grit and perseverance alone won't save you from terrible diseases. If that were the case, then so many of those who have died would be alive today. No, what our experience as HIV researchers and survivors of a superbug experience shows us is that when scientific advances, masterful medicine, and the will to live come together, that's when the impossible becomes possible.

TO READERS

The information in this book is not a substitute for professional medical attention. For more resources, check out https://IPATH.UCSD.EDU and ThePerfectPredator.com.

ACKNOWLEDGMENTS

We succeeded in saving Tom's life, but thanking all the people who helped do that—and those who supported us in writing this book—is a daunting task. Healing and helping hands reached across the globe, sending cards and messages, lighting candles, bringing meals and care packages, hunting phages, transporting samples, sitting bedside, donating blood, praying, and of course, caring for him and our family. In some cases, we don't even know their names.

We owe a debt of gratitude to the staff from the clinic in Luxor, our guide Khalid, the twenty-four-hour medical staff at our travel insurance company, United Health Care Services Inc., and the medevac teams and the ICU at Frankfurt's Uniklinik, who kept Tom's condition stable enough so he could be transported safely back to San Diego.

The medical and nursing team at the Thornton ICU and hospital in La Jolla (now the Jacobs Medical Center) are our heroes, especially Chip Schooley and Davey Smith. From the bottom of our hearts, we thank faculty, staff, students, fellows, and leadership at UCSD Health Sciences, especially those in the Department of Medicine, including faculty, staff, fellows, and students in the Divisions of Pulmonary and Critical Care, Infectious Disease and Global Health, and the Antiviral Research Center. We regret that we could only name a few in this book.

Huge thanks to Ry Young, Jason Gill, Adriana Hernandez-Morales, Jacob Lancaster and others on the CPT team at Texas A&M, and to the US Navy BDRD, specifically Lt. Commander Theron Hamilton, Biswajit Biswas, Captains Mateczun and Stockelman, Lt. Commander Luis Estrella, Matthew Henry, and Javier Quinones, for their tireless work to generate phage cocktails for Tom. An unsung hero is Dr. Cara Fiore from the FDA, who went beyond the call of duty to connect us to phage researchers and to fast-track the necessary approvals. We also thank the leadership from AmpliPhi Biosciences for donating a phage, and researchers from India, Switzerland, and Belgium who offered phages as well. Forest Rohwer, Anca Segall, and their labs at San Diego State University, especially visiting postdoc Jeremy Barr, and PhD student Sean Benler, who saved the day by stepping in to repurify the first phage cocktail on a moment's notice. Carl Merril and Maia Merabishvili provided critical advice on phage dosing, administration, and safety, and Charles Dinarello on endotoxin levels.

Our families played a critical role in supporting us in so many ways. My dad, Al Keith, died before this book was complete. After reading an early draft, one of his last words to us was "Bravo!" Our daughters Carly and Frances, and their mom Suzi; son Cameron; and Danny were lifelines. My mom, Heather; sister Jill and her daughters, Rylie and Morgan; and my sister Jennifer, her husband, Pete, and their kids, Ella and Nathan, were always there for us.

We are indebted to Martin Feisst and Robert Lindsy Milne for their energy, intuitive advice, and support. Countless friends, students, postdocs, faculty, and staff who are too numerous to name kept our spirits from flagging with visits, dinners, phone calls, and running errands, which meant the world to us. Friends on Facebook and Twitter gave us the encouragement we needed to work on this book, even when the memories were painful.

Special thanks to our friend Jon Cohen and his wife, Shannon Bradley, who not only visited Tom and nourished our family repeatedly, but who also provided helpful advice that helped us find both an amazing agent and a publisher.

In writing this book, we stayed as close to the true story as possible, referring to fifty-two pages of my Facebook entries, over three thousand pages of Tom's medical records, scads of email threads, and conversations with those involved when memories got fuzzy. In a few cases, we changed names to protect people's privacy, but in most cases, real names were used.

We are grateful to many individuals who repeatedly checked drafts of this book for accuracy. Chip Schooley and Davey Smith checked medical details and descriptions, and Liz Greer painstakingly corrected several drafts, even after losing her husband, Doug, to pancreatic cancer not long after Tom came home. Carly and Frances Patterson, Cameron Strathdee, Trish Case, Joe DeSommer, Judy Auerbach, Diana McCague, Steve Weiner, Kristen Rau, and Margaret Browning provided advice on readability. Hubert Mazure, Bill Summers, Betty Kutter, Gordon Wheat, Bob Blasdel, Ry Young, Biswajit Biswas, Theron Hamilton, Jason Gill, and Carl Merril provided clarifications on historical and scientific details. Brian Kelly sought attributions for quotes and obtained necessary permissions. Dr. Charles Pope provided a scanning electron micrograph of Tom's *Acinetobacter* being attacked by Navy phages. (We have an enlarged copy we throw darts at.)

This book would not have been published if it wasn't for our agents, Gail Ross and Dara Kaye at Ross-Yoon, who believed in us and saw us through many proposal drafts. We are equally grateful to our executive editors, Michelle Howry, Amanda Murray, and Krishan Trotman; copy editor Lori Paximadis; art director Amanda Kain; our publicist, Joanna Pinsker; and our publisher, Mauro DiPreta from Hachette Book Group, for their unwavering enthusiasm, support, and wisdom. We also thank Ilsa Brink for our terrific book website: ThePerfectPredator.com.

Steff would like to acknowledge the role of several scientists who were pivotal mentors in her early career: Michael Sukhdeo, Randall Coates, Stanley Read, and Michael V. O'Shaughnessy. She also thanks the anonymous family donors of her Harold Simon endowed chair that has supported her research at UCSD.

Our co-writer, Teresa Barker, was more than a partner in our journey. She relived our experiences, shared in our tears and laughter, and ultimately helped us hone our story so it would speak to everyone. Teresa, you are family now. Her husband, Steve Weiner, was of inestimable support, as were Dolly Joern, Kristen and Rachel Rau, Becca Barker, Aaron and Lauren Weiner, Margaret Browning, Sue Shellenbarger, Leslie

Rowan, Elizabeth Leibowitz, Wendy Miller, and others who supported the project in different ways. Teresa's young grandchildren, Leyna and Aden, were personal reminders of the world of children at stake, whose lives depend on the work of those in science, medicine, and health policy to protect them from the burgeoning superbug pandemic.

We also want to acknowledge the special role of Mallory's parents, Diane Shader-Smith and Mark Smith, who provided us with endless encouragement as we wrote our book. Even though phages arrived too late to save their daughter Mallory, they pledged the inaugural gift to IPATH, our new phage therapy center at UC San Diego. I was deeply honored that they asked me to write the epilogue in Mallory's memoir, *Salt in My Soul: An Unfinished Life*.

We are fortunate to be alive at a time when science and technology have advanced to a point that the zeitgeist made Tom's cure not just a theoretical possibility but a reality. Without the scores of people who took incalculable risks and devoted time and resources to our struggle, this story would have been an ordinary death, and one of an estimated 1.5 million people who lose their lives to superbugs every year. Our hope is that this book will increase awareness of the growing superbug crisis and propel more research on phage therapy. Knowing that our experience has begun to help others makes the pain and suffering our family endured worthwhile. And for those of you out there who are battling serious superbug infections, we are with you. *No retreat.*

SELECTED CHAPTER REFERENCES

Chapter 3. ***Disease Detectives***

"Foodborne Illnesses and Germs." Centers for Disease Control and Prevention, https://www.cdc.gov/foodsafety/foodborne-germs.html.

Johnson, Steven. *The Ghost Map: The Story of London's Most Terrifying Epidemic—and How It Changed Science, Cities, and the Modern World*. New York: Riverhead Books, 2006.

Chapter 4. ***First Responders***

McKenna, Maryn. *Superbug: The Fatal Menace of MRSA*. New York: Free Press, 2011.

Chapter 5. ***Lost in Translation***

Lax, Eric. *The Mold in Dr. Florey's Coat: The Story of the Penicillin Miracle*. New York: Henry Holt, 2004.

Chapter 7. ***A Deadly Hitchhiker***

Lankisch, P. G., M. Apte, and P. A. Banks. "Acute Pancreatitis." *Lancet* 386, no. 9988 (July 4, 2015): 85–96.

Stinton, L. M., R. P. Myers, and E. A. Shaffer. "Epidemiology of Gallstones." *Gastroenterology Clinics of North America* 39, no. 2 (June 2010): 157–169, vii.

Chapter 8. ***"The Worst Bacteria on the Planet"***

Boucher, Helen W., George H. Talbot, John S. Bradley, John E. Edwards, David Gilbert, Louis B. Rice, Michael Scheld, Brad Spellberg, and John Bartlett. "Bad Bugs, No Drugs: No Eskape! An Update from the Infectious Diseases Society of America." *Clinical Infectious Diseases* 48, no. 1 (2009): 1–12.

Camp, Callie, and Owatha L. Tatum. "A Review of *Acinetobacter baumannii* as a Highly Successful Pathogen in Times of War." *Laboratory Medicine* 41, no. 11 (2010): 649–657.

Rice, L. B. "Federal Funding for the Study of Antimicrobial Resistance in Nosocomial Pathogens: No Eskape." *Journal of Infectious Diseases* 197, no. 8 (April 15, 2008): 1079–1081.

Silberman, Steve. "The Invisible Enemy." *Wired*, February 1, 2007. https://www.wired.com/2007/02/enemy.

Wong, D., T. B. Nielsen, R. A. Bonomo, P. Pantapalangkoor, B. Luna, and B. Spellberg. "Clinical and Pathophysiological Overview of Acinetobacter Infections: A Century of Challenges." *Clinical Microbiology Reviews* 30, no. 1 (January 2017): 409–447.

Chapter 11. *Public Enemy No. 1: Under the Radar*

Blaser, Martin J. *Missing Microbes: How the Overuse of Antibiotics Is Fueling Our Modern Plagues*. New York: Picador, 2015.

Chen, L., R. Todd, J. Kiehlbauch, M. Walters, and A. Kallen. "Notes from the Field: Pan-Resistant New Delhi Metallo-Beta-Lactamase-Producing Klebsiella Pneumoniae—Washoe County, Nevada, 2016." *Morbidity and Mortality Weekly Report* 66, no. 1 (January 13, 2017): 33.

Cohen B., S. Hyman, L. Rosenberg, and E. Larson. "Frequency of Patient Contact with Health Care Personnel and Visitors: Implications for Infection Prevention." *Joint Commission Journal on Quality and Patient Safety/Joint Commission Resources* 38, no. 12 (2012): 560–565. https://www.ncbi.nlm.nih.gov/pmc/articles/PMC3531228.

Doyle, J. S., K. L. Buising, K. A. Thursky, L. J. Worth, and M. J. Richards. "Epidemiology of Infections Acquired in Intensive Care Units." *Seminars in Respiratory and Critical Care Medicine* 32, no. 2 (April 2011): 115–138.

"Global Priority List of Antibiotic-Resistant Bacteria to Guide Research, Discovery, and Development of New Antibiotics," World Health Organization. http://www.who.int/medicines/publications/global-priority-list-antibiotic-resistant-bacteria/en.

Huslage, K., et al. "A Quantitative Approach to Defining 'High-Touch' Surfaces in Hospitals." *Infection Control and Hospital Epidemiology* 31, no. 8 (2010): 850–853.

Lax, S., N. Sangwan, D. Smith, P. Larsen, K. M. Handley, M. Richardson, K. Guyton, M. Krezalek, B. D. Shogan, J. Defazio, I. Flemming, B. Shakhsheer, S. Weber, E. Landon, S. Garcia-Houchins, J. Siegel, J. Alverdy, R. Knight, B. Stephens, and J. A. Gilbert. "Bacterial Colonization and Succession in a Newly Opened Hospital." *Science Translational Medicine* 9 (May 24, 2017).

Laxminarayan, R., and R. R. Chaudhury. "Antibiotic Resistance in India: Drivers and Opportunities for Action." *PLoS Medicine* 13, no. 3 (March 2016): e1001974.

Liu, Cindy M., M. Stegger, M. Aziz, T. J. Johnson, K. Waits, L. Nordstrom, L. Gauld, B. Weaver, D. Rolland, S. Statham, J. Horwinski, S. Sariya, G. S. Davis, E. Sokurenko, P. Keim, J. R. Johnson, and L. B. Price. "*Escherichia coli* ST131-*H*22 as a Foodborne Uropathogen." *mBio* 9, no. 4 (August 2018); DOI: 10.1128/mBio.00470-18.

McKenna, Maryn. *Big Chicken: The Incredible Story of How Antibiotics Created Modern Agriculture and Changed the Way the World Eats*. New York: Penguin, 2017.

Ofstead, C. L., H. P. Wetzler, E. M. Doyle, C. K. Rocco, K. H. Visrodia, T. H. Baron, and P. K. Tosh. "Persistent Contamination on Colonoscopes and Gastroscopes Detected by Biologic Cultures and Rapid Indicators Despite Reprocessing Performed in Accordance with Guidelines." *American Journal of Infection Control* 43, no. 8 (August 2015): 794–801.

Terhune, Chad. "Olympus Told Its US. Executives No Broad Warning about Tainted Medical Scopes Was Needed, Despite Superbug Outbreaks." *Los Angeles Times*, July 21, 2016. http://www.latimes.com/business/la-fi-olympus-scopes-emails-20160721-snap-story.html.

Chapter 12. *The Alternate Reality Club*

Gelling, L. "Causes of ICU Psychosis: The Environmental Factors." *Nursing and Critical Care* 4, no. 1 (January–February 1999): 22–26.

Lin, L., P. Nonejuie, J. Munguia, A. Hollands, J. Olson, Q. Dam, M. Kumaraswamy, et al. "Azithromycin Synergizes with Cationic Antimicrobial Peptides to Exert Bactericidal and Therapeutic Activity against Highly Multidrug-Resistant Gram-Negative Bacterial Pathogens." *EBioMedicine* 2, no. 7 (July 2015): 690–698.

Chapter 13. *Tipping Point: Fully Colonized*

Burnham, J., Olsen, M., & Kollef, M. (n.d.). "Re-estimating Annual Deaths Due to Multidrug-Resistant Organism Infections." *Infection Control & Hospital Epidemiology*, 1–2. doi:10.1017/ice.2018.304.

Chan, Margaret. "Antimicrobial Resistance in the European Union and the World." World Health Organization, 2012. http://www.who.int/dg/speeches/2012/amr_20120314/en.

Liu, Y. Y., Y. Wang, T. R. Walsh, L. X. Yi, R. Zhang, J. Spencer, Y. Doi, et al. "Emergence of Plasmid-Mediated Colistin Resistance Mechanism Mcr-1 in Animals and Human Beings in China: A Microbiological and Molecular Biological Study." *Lancet Infectious Disease* 16, no. 2 (February 2016): 161–168.

Seymour, C. W., and M. R. Rosengart. "Septic Shock: Advances in Diagnosis and Treatment." *Journal of the American Medical Association* 314, no. 7 (August 18, 2015): 708–717.

"Tackling Drug-Resistant Infections Globally: Final Report and Recommendations." May 2016. Review on Antimicrobial Resistance, commissioned by Her Majesty's Government (UK) and the Wellcome Trust. https://amr-review.org/sites/default/files/160525_Final%20paper_with%20cover.pdf.

Walsh, T. R., and Y. Wu. "China Bans Colistin as a Feed Additive for Animals." *Lancet Infectious Disease* 16, no. 10 (October 2016): 1102–1103.

Chapter 14. *The Spider to Catch the Fly*

Fishbain, J., and A. Y. Peleg. "Treatment of *Acinetobacter* Infections." *Clinical Infectious Disease* 51, no. 1 (July 1, 2010): 79–84.

Garcia-Quintanilla, M., M. R. Pulido, R. Lopez-Rojas, J. Pachon, and M. J. McConnell. "Emerging Therapies for Multidrug Resistant *Acinetobacter Baumannii*." *Trends in Microbiology* 21, no. 3 (March 2013): 157–163.

Geoghegan, J. L., and E. C. Holmes. "Predicting Virus Emergence amid Evolutionary Noise." *Open Biology* 7, no. 10 (October 2017).

Ghorayshi, Azeen. "Mail-Order Viruses Are the New Antibiotics." *BuzzFeed*, February 2, 2015. https://www.buzzfeed.com/azeenghorayshi/mail-order-viruses-are-the-new-antibiotics.

Hendrickson, Heather. "Nature's Ninjas in the Battle against Superbugs." TED Talk, October 6, 2016. https://www.youtube.com/watch?v=p2ngpKBPfF8.

Kuchment, Anna. *The Forgotten Cure: The Past and Future of Phage Therapy*. New York: Springer, 2012.

Merabishvili, M., D. Vandenheuvel, A. M. Kropinski, J. Mast, D. De Vos, G. Verbeken, J. P. Noben, et al. "Characterization of Newly Isolated Lytic Bacteriophages Active against *Acinetobacter baumannii*." *PLoS One* 9, no. 8 (2014): e104853.

Mokili, J., Rohwer, F., Dutih, B. E. "Metagenomics and Future Perspectives in Virus Discovery."*Current Opinion in Virology* 2, no. 1 (February 2012): 63–77. https://www.sciencedirect.com/science/article/pii/S1879625711001908?via%3Dihub.

Chapter 15. *The Perfect Predator*

d'Hérelle, Félix. *The Bacteriophage, Its Rôle in Immunity.* Toronto: University of Toronto, 1922.

Doudna, Jennifer, and Samuel Sternberg. *A Crack in Creation: Gene Editing and the Unthinkable Power to Control Evolution.* Boston: Mariner Books, 2017.

Merril, C. R., B. Biswas, R. Carlton, N. C. Jensen, G. J. Creed, S. Zullo, and S. Adhya. "Long-Circulating Bacteriophage as Antibacterial Agents." *Proceedings of the National Academy of Sciences USA.* 93, no. 8 (April 16, 1996): 3188–3192.

Reardon, Sara. "Phage Therapy Gets Revitalized." *Nature* 510, no. 7503 (June 4, 2014). https://www.nature.com/news/phage-therapy-gets-revitalized-1.15348.

Summers, W. C. "Bacteriophage Therapy." *Annual Review of Microbiology* 55 (2001): 437–451.

Summers, W. C. "Félix Hubert d'Hérelle (1873–1949): History of a Scientific Mind." *Bacteriophage* 6, no. 4 (2016): e1270090.

Chapter 18. *Panning for Gold*

Henry, M., B. Biswas, L. Vincent, V. Mokashi, R. Schuch, K. A. Bishop-Lilly, and S. Sozhamannan. "Development of a High Throughput Assay for Indirectly Measuring Phage Growth Using the Omnilog™ System." *Bacteriophage* 2, no. 3 (July 1, 2012): 159–167.

Kutter, E. M., S. J. Kuhl, and S. T. Abedon. "Re-Establishing a Place for Phage Therapy in Western Medicine." *Future Microbiology* 10, no. 5 (2015): 685–688.

Merril, C. R., D. Scholl, and S. L. Adhya. "The Prospect for Bacteriophage Therapy in Western Medicine." *National Review of Drug Discovery* 2, no. 6 (Jun 2003): 489–497.

Pirnay, J. P., D. De Vos, G. Verbeken, M. Merabishvili, N. Chanishvili, M. Vaneechoutte, M. Zizi, et al. "The Phage Therapy Paradigm: Prêt-à-Porter or Sur-Mesure?" *Pharmaceutical Research* 28, no. 4 (April 2011): 934–937.

Snitkin, E. S., A. M. Zelazny, P. J. Thomas, F. Stock, Nisc Comparative Sequencing Program Group, D. K. Henderson, T. N. Palmore, and J. A. Segre. "Tracking a Hospital Outbreak of Carbapenem-Resistant *Klebsiella pneumoniae* with Whole-Genome Sequencing." *Science Translational Medicine* 4, no. 148 (August 22, 2012): 148ra16.

Young, R., and J. J. Gill. "Microbiology. Phage Therapy Redux—What Is to Be Done?" *Science* 350, no. 6265 (December 4, 2015): 1163–1164.

Chapter 20. *The Blood Orange Tree*

Keller, Evelyn Fox. *A Feeling for the Organism, 10th Anniversary Edition: The Life and Work of Barbara McClintock.* New York: Henry Holt, 1983.

Chapter 21. *Moment of Truth*

Bhargava, N., P. Sharma, and N. Capalash. "Quorum Sensing in *Acinetobacter*: An Emerging Pathogen." *Critical Reviews in Microbiology* 36, no. 4 (November 2010): 349–360.

Borges, A. L., J. Y. Zhang, M. F. Rollins, B. A. Osuna, B. Wiedenheft, and J. Bondy-Denomy. "Bacteriophage Cooperation Suppresses CRISPR-Cas3 and Cas9 Immunity." *Cell* 174, no. 4 (August 9, 2018): 917–925.e10.

Erez, Z., I. Steinberger-Levy, M. Shamir, S. Doron, A. Stokar-Avihail, Y. Peleg, S. Melamed, et al. "Communication between Viruses Guides Lysis-Lysogeny Decisions." *Nature* 541, no. 7638 (January 26, 2017): 488–493.

Harding, C. M., S. W. Hennon, and M. F. Feldman. "Uncovering the Mechanisms of *Acinetobacter baumannii* Virulence." *National Review of Microbiology* 16, no. 2 (February 2018): 91–102.

Logan, L. K., S. Gandra, A. Trett, R. A. Weinstein, and R. Laxminarayan. "*Acinetobacter baumannii* Resistance Trends in Children in the United States, 1999–2012." *Journal of the Pediatric Infectious Disease Society* (March 22, 2018).

Young, R. "Phage Lysis: Three Steps, Three Choices, One Outcome." *Journal of Microbiology* 52, no. 3 (March 2014): 243–258.

Chapter 22: *The Bold Guess*

Meldrum, M. "'A Calculated Risk': The Salk Polio Vaccine Field Trials of 1954." *British Medical Journal* 317, no. 7167 (October 31, 1998): 1233–1236.

Nguyen, S., K. Baker, B. S. Padman, R. Patwa, R. A. Dunstan, T. A. Weston, K. Schlosser, B. Bailey, T. Lithgow, M. Lazarou, A. Luque, R. Rohwer, R. S. Blumberg, and J. J. Barr. "Bacteriophage Transcytosis Provides a Mechanism to Cross Epithelial Cell Layers." *MBio* 8, no. 6 (November 21, 2017).

Chapter 24. *Second-Guessing*

Sacks, Oliver. *Awakenings*. New York: Vintage, 1999.

Chapter 26. *The Darwinian Dance and the Red Queen's Pursuit*

Brockhurst, M. A., T. Chapman, K. C. King, J. E. Mank, S. Paterson, and G. D. Hurst. "Running with the Red Queen: The Role of Biotic Conflicts in Evolution." *Proceedings of the Royal Society B: Biological Sciences* 281, no. 1797 (December 22, 2014).

Regeimbal, J. M., A. C. Jacobs, B. W. Corey, M. S. Henry, M. G. Thompson, R. L. Pavlicek, J. Quinones, R. M. Hannah, M. Ghebremedhin, N. J. Crane, D. V. Zurawski, N. C. Teneza-Mora, B. Biswas, and E. R. Hall. "Personalized Therapeutic Cocktail of Wild Environmental Phages Rescues Mice from *Acinetobacter baumannii* Wound Infections." *Antimicrobial Agents and Chemotherapy* 60, no. 10 (October 2016).

Scholl, D., J. Kieleczawa, P. Kemp, J. Rush, C. C. Richardson, C. Merril, S. Adhya, and I. J. Molineux. "Genomic Analysis of Bacteriophages Sp6 and K1-5, an Estranged Subgroup of the T7 Supergroup." *Journal of Molecular Biology* 335, no. 5 (January 30, 2004): 1151–1171.

Chapter 27. *The Last Dance*

Rohwer, Forest, Heather Maughan, Merry Youle, and Nao Hisakawa. *Life in Our Phage World: A Centennial Field Guide to the Earth's Most Diverse Inhabitants*. San Diego: Wholon, 2014.

Summers, W. C. "The Strange History of Phage Therapy." *Bacteriophage* 2, no. 2 (April 1, 2012): 130–133.

Chapter 28. *The Buddha's Gift*

Mukherjee, Siddhartha. "The Rules of the Doctor's Heart." *New York Times*, October 24, 2017. https://www.nytimes.com/2017/10/24/magazine/the-rules-of-the-doctors-heart.html.

Chapter 29. *Grand Rounds*

BMJ 2018, 363 doi: https://doi.org/10.1136/bmj.k4762. (Published 08 November 2018.

CDC Telebriefing on Today's Drug-resistant Health threats. https://www.cdc.gov/media/releases/2013/t0916_health-threats.html.

Chan, B. K., P. E. Turner, S. Kim, H. R. Mojibian, J. A. Elefteriades, and D. Narayan. "Phage Treatment of an Aortic Graft Infected with *Pseudomonas aeruginosa*." *Evolution, Medicine, and Public Health* 2018, no. 1 (2018): 60–66.

Davies, Madlen. "A Game of Chicken: How Indian Poultry Farming Is Creating Global Superbugs." Bureau of Investigative Journalism. January 2018. https://www.thebureauinvestigates.com/stories/2018-01-30/a-game-of-chicken-how-indian-poultry-farming-is-creating-global-superbugs.

Garrett, L., and R. Laxminarayan "Antibiotic-Resistant 'Superbugs' Are Here." https://www.cfr.org/expert-brief/antibiotic-resistant-superbugs-are-here.

Hall, William. *Superbugs: An Arms Race against Bacteria*. Cambridge, MA: Harvard University Press, 2018.

"High-Level Meeting on Antimicrobial Resistance." United Nations General Assembly, September 2016. https://www.un.org/pga/71/event-latest/high-level-meeting-on-antimicrobial-resistance.

Jennes, S., M. Merabishvili, P. Soentjens, K. W. Pang, T. Rose, E. Keersebilck, O. Soete, et al. "Use of Bacteriophages in the Treatment of Colistin-Only-Sensitive *Pseudomonas aeruginosa* Septicaemia in a Patient with Acute Kidney Injury—a Case Report." *Critical Care* 21, no. 1 (June 4, 2017): 129.

Lyon, J. "Phage Therapy's Role in Combating Antibiotic-Resistant Pathogens." *Journal of the American Medical Association* 318, no. 18 (November 14, 2017): 1746–1748.

OECD. "Stemming the Superbug Tide: Just a Few Dollars More." *OECD Health Policy Studies*. Paris: OECD Publishing, 2018. https://doi.org/10.1787/9789264307599-en.

Schooley, R. T., B. Biswas, J. J. Gill, A. Hernandez-Morales, J. Lancaster, L. Lessor, J. J. Barr, et al. "Development and Use of Personalized Bacteriophage-Based Therapeutic Cocktails to Treat a Patient with a Disseminated Resistant *Acinetobacter baumannii* Infection." *Antimicrobial Agents and Chemotherapy* 61, no. 10 (October 2017).

Servick, K. "U.S. Center Will Fight Infections with Viruses." *Science* 360, no. 6395 (June 22, 2018): 1280–1281.

Stockton, Ben. "Antibiotics in Agriculture: The Blurred Line between Growth Promotion and Disease and Prevention." Bureau of Investigative Journalism. September 2018. https://www.thebureauinvestigates.com/stories/2018-09-19/growth-promotion-or-disease-prevention-the-loophole-in-us-antibiotic-regulations.

Stockton, B., Davies, M., Meesaraganda, R. "Zoetis and Its Antibiotics for Growth in India." *Veterinary Record* 183 (October 2018), 432–433. https://veterinaryrecord.bmj.com/content/183/14/432.

Watts, G. "Phage Therapy: Revival of the Bygone Antimicrobial." *Lancet* 390, no. 10112 (December 9, 2017): 2539–2540.

Epilogue

Davidson, J. E., K. Powers, K. M. Hedayat, M. Tieszen, A. A. Kon, E. Shepard, V. Spuhler, et al. "Clinical Practice Guidelines for Support of the Family in the Patient-Centered Intensive Care Unit: American College of Critical Care Medicine Task Force 2004–2005." *Critical Care Medicine* 35, no. 2 (February 2007): 605–622.

Davidson, J. E., and S. A. Strathdee. "The Future of Family-Centred Care in Intensive Care." *Intensive and Critical Care Nursing* (March 29, 2018).

Davydow, D. S., J. M. Gifford, S. V. Desai, D. M. Needham, and O. J. Bienvenu. "Posttraumatic Stress Disorder in General Intensive Care Unit Survivors: A Systematic Review." *General Hospital Psychiatry* 30, no. 5 (September–October 2008): 421–434.

"Monitoring Global Progress on Addressing Antimicrobial Resistance: Analysis Report of the Second Round of Results of AMR Country Self-Assessment Survey 2018." http://apps.who.int/iris/bitstream/handle/10665/273128/9789241514422-eng.pdf.

Palms, D. L., L. A. Hicks, M. Bartoces, et al. "Comparison of Antibiotic Prescribing in Retail Clinics, Urgent Care Centers, Emergency Departments, and Traditional Ambulatory Care Settings in the United States." *JAMA Internal Medicine* 178, no. 9 (2018): 1267–1269. doi:10.1001/jamainternmed.2018.1632.

"Tracking Progress to Address AMR." AMR Industry Alliance. January 2018. https://www.amrindustryalliance.org/progress-report.

ABOUT THE AUTHORS

Steffanie Strathdee, PhD

Dr. Steffanie Strathdee obtained her BSc, MSc, and PhD from the University of Toronto. A dual citizen of Canada and the US, she is associate dean of global health sciences, professor and Harold Simon Chair at the University of California, San Diego, School of Medicine. She also co-directs UC San Diego's Center for Innovative Phage Applications and Therapeutics (IPATH) and UC San Diego Global Health Institute and is an adjunct professor at Johns Hopkins and Simon Fraser Universities. Steffanie is an infectious disease epidemiologist who has nearly twenty years' experience and over six hundred peer-reviewed publications on the prevention of HIV, sexually transmitted infection, and viral hepatitis among marginalized populations. In 2018, she was one of *Time*'s Most Influential People in Health Care.

Thomas Patterson, PhD

Dr. Tom Patterson obtained his AB from San Diego State University, his MSc from the University of Georgia, and his PhD from UC Riverside. Tom is an evolutionary sociobiologist and an experimental psychologist. He is a distinguished professor of psychiatry at UC San Diego and has renowned expertise on behavioral interventions among HIV-positive persons and those at high risk of acquiring HIV and sexually transmitted infections. He developed a scale to assess everyday functioning in schizophrenia that has been mandated by the FDA and is in wide use, having been translated into twenty-six languages.

Drs. Patterson and Strathdee have worked as a husband-and-wife research team on the Mexico–US border for over a decade. Their work inspired Jon Cohen's book *Tomorrow Is a Long Time*, published in 2015. This is their first book together.

Teresa H. Barker, co-writer

A career journalist and book collaborator, Teresa Barker has co-written more than a dozen published titles in the fields of health, parenting and child development, spirituality, and creativity and aging, including the *New York Times* bestseller *Raising Cain: Protecting the Emotional Life of Boys* with Michael Thompson, PhD; *The Soul of Money: Transforming Your Relationship with Money and Life* with Lynne Twist, and *The Big Disconnect: Protecting Childhood and Family Relationships in the Digital Age* with Catherine Steiner-Adair, EdD, which was named by the *Wall Street Journal* to the Top Ten Nonfiction Books of 2013.

INDEX

Note: In this index the following abbreviations are used: TP for Tom Patterson, and SS for Steffanie Strathdee